"America's Public Enemy #1 today is our addiction to our smartphones and tablets. They are distorting our collective sense of reality. *Connected & Engaged* is a fascinating read! Dr. Whatley gives refreshing and thought-provoking advice on how we can live our best lives in this age of technology."

KAEDY KIELY,
Afternoon Drive Host at WSRV-FM,
Cox Media Group, Atlanta

"Lori Whatley skillfully shows us how technology has become a form of addictive brain sugar that distracts us from real connection and steals our joy. We can never get back the time we spend deleting spam. This book helped me use technology to increase productivity, reduce stress, improve my relationships, and feed my brain with the same intention I feed my body."

KRISTEN COFFIELD,
Author, Educator and Culinary Disruptor, Founder and
CEO of The Culinary Cure, TheCulinaryCure.com

"Lori Whatley has given us the thing we all want the most— meaningful connections with the people around us. This book is a powerful, yet simple blueprint to help us manage our distractions and focus on what matters most, the people in our lives."

ANTON J. GUNN,
MSW, CDM, CSP, International Speaker &
Leadership Consultant, *Helping Leaders Build*
World-Class Culture, AntonGunn.com

"An honest look at our digital lives with real-life solutions to help us personally refresh, recharge and reconnect with the people we value off screen."

KELLY JAMESON,
PhD, LPC-S, Therapy. Speaking.
Consulting. DrKellyJameson.com

"We can't multiply our time when we're distracted by digital devices. In this lively read, Lori Whatley equips us with practical suggestions to help us put people first."

RORY VADEN,
New York Times bestselling author of Take the Stairs

CONNECTED
AND
ENGAGED

How to Manage
Digital Distractions and Reconnect
with the World around You

DR. LORI WHATLEY

Hatherleigh Press is committed to preserving and protecting the natural resources of the earth. Environmentally responsible and sustainable practices are embraced within the company's mission statement.

Visit us at www.hatherleighpress.com.

CONNECTED & ENGAGED

Library of Congress Cataloging-in-Publication Data is available upon request.
ISBN: 978-1-961293-06-9. 2nd edition.

Cover design by Carolyn Kasper
Author photograph by Sarah Turner Taboada

Printed in the United States
10 9 8 7 6 5 4 3 2 1

For my darling granddaughter Ann Warren. For you, I want to positively influence the world and make it the very best place for your generation to enjoy emotional balance and well-being.

Never doubt that a small group of thoughtful, committed citizens can change the world; indeed, it's the only thing that ever has.

—MARGARET MEAD

CONTENTS

AUTHOR'S NOTE

MANY YEARS AGO, I foresaw a dark cloud gathering on the horizon—hovering over my therapy sessions like a grim portent of things to come.

The storm that rolled into my office came in many different forms. One day, it would look like a couple in a heated argument because one spouse preferred screen time over couple time. The next day, it would be a teenager failing school, chronically exhausted from late nights gaming on his digital devices. The day after that, I'd see the CEO of a Fortune 500 company who was so deep into online pornography and the addiction it birthed that he simply couldn't find a way out. His marriage was failing, his workload was suffering and his life was spinning out of control in almost every area. He was deeply depressed.

In almost every generation, I saw people struggling with the challenges of digital device usage. I decided to write a book about the pitfalls of careless digital device usage—to share with the world a warning about this lurking darkness and the danger it posed to the unassuming human. No one appeared to be talking about it and it seemed like a conversation that needed to begin.

The same week that *Connected & Engaged* was first published, the world was being consumed by the start of the COVID-19 pandemic. As a result, I had to cancel my in-person book tour, aimed at introducing my new ideas about harmful social media and excessive digital device usage to the world. It would have to wait, I thought.

Weeks turned into months, as social isolation continued to blanket our world. All the while, the dark cloud I'd first seen continued to grow, becoming a sister pandemic to COVID. I'm sure we can all recall the days and nights spent alone, scrolling. Screens were everything for us in this new quarantined society—our only form of connection with the world outside the four walls we were all confined within.

We learned to live our lives from our laptops. We adapted to work, socialize and so much more, all through a touchphone's screen. But when the quarantine finally lifted and we were able to return to in-person interactions, these new habits didn't just go away. For some, the effects of the constant online exposure seen during COVID have created a new normal. A post-COVID normal. An isolated, deleterious normal.

The storm clouds are now bigger than ever. Having gathered speed and strength through an unprecedented global pandemic, they are currently wreaking havoc in our world in ways we could never have imagined. Simply put, our world has become addicted to our digital devices…and like any addiction, there is a deep personal and social cost. The loss of community has resulted in deserted marriages, abandoned in-person friendships, and the complete absence of peace of mind. Anxiety and depression have grown by leaps and bounds in the younger generations, who spend hours and hours head down, fingers scrolling, oblivious to the connections that are built when we look into another human's eyes.

Connected & Engaged is more relevant now than ever. We have lost the art of connection, we have lost our way, and we are suffering for it. But the great news is that we *can* regain our path toward community. We learn new ways (and bring back some of the old) of conversing that will bring us improved mental and physical health.

The surgeon general recently issued an urgent call for actions to educate our world about the negative impact of social media on our youth and the importance of connecting in person and of building in-person community experiences. He has challenged policymakers and technology companies to take urgent action to turn this epidemic around and make our world safe again—not only for our kids, but for everyone. It's not enough just to understand the negative impact of excessive screen usage—though that *is* an important first step. We must also implement healthier practices, like those mentioned in this book, to bring balance into our lives.

At present, we have much deeper insight into how our minds are affected by excessive digital device usage. We also have information about the habits needed to heal us and reconnect with our mental and physical health. For example: we understand that texting is useful for a quick check-in but that it's not a healthy way to carry on a relationship. In other words, there exists a fine line between healthy texting and misuse.

Every individual—and especially all parents—must make it a top priority to educate themselves on safe digital device usage. I have found it immensely rewarding to share with educators and others the importance of striking a good tech-life balance and making wise choices around digital device usage. However, the changes needed must begin in our homes and workplaces.

This is a big job and it will take us all to relearn how to identify what benefits our community and what harms our personal connection. Knowledge is power. We can do this!

I'll end with this thought. We are all works in progress. We are all on a journey and each new day is a part of that journey, bringing with it new opportunities to make healthier choices. Today, choose real, in-person connection. Limit your screen time just for one day and take notice of the many benefits it brings. You *can* open the door to unlimited possibilities, one good choice at a time.

Read on and find out the many options available to you for a healthier, more connected and engaged life.

INTRODUCTION

I am interested in real people, real stories and real connection.

—Jamie Lee Curtis

Jenny came to my office like many clients for their first session—depressed, confused, and hurting. Jenny had been married for three years, and the connection with her partner was becoming less than good. Jenny could not put her finger on the problem. She only knew her husband seemed distracted and distant.

He was often on his phone texting and stayed up far later than Jenny; he said he was "working" on his laptop. He was irritable when he was too far from his digital devices and alone time for them had become non-existent. She felt like her husband's mind had been hijacked, and talking to him about this was like talking to a brick wall.

All she got was a blank stare from empty eyes. Being in this marriage was becoming more and more lonely. She had not contemplated competing with digital devices when she agreed to "happily ever after" with Tom. Now, he had no focus on the relationship, and she was losing hope that he ever would.

Her husband had become like millions of others who have traded in their real-life *love* for online *likes*. Little did Tom know, the two were not the same at all.

Here's the hard truth: we carry around in our pockets digital slot machines, and we are addicted. They constantly battle for our attention, and they distract us from the real connections which we need to thrive.

We check our phones over 150 times per day and spend 10 hours and 39 minutes on average each day in front of a screen. That is more than 2/3 of our waking hours! This affects our mental health, and consequently, our relationships are suffering. We now have a shorter attention span than our pet goldfish due to these distractions.

Our children are facing new obstacles that other generations have not had to navigate. They are handicapped when socializing due to their constant interaction with technology. Many people who work for tech companies and design technology do not allow their kids to use technology. They limit their kids' time online, yet the general population continues to allow kids to take risks involved with excessive tech use.

Tech companies have built-in intermittent rewards and dopamine-driven feedback loops that are obliterating our society. The like button on our digital devices is driving many of our lives. We are unable to be truly present with one another because we can't disconnect from our digital devices and truly connect and engage as we were created to do. Apps are created to be addictive. They keep us more focused on our digital devices and online longer, which unplugs us from our real-life experiences.

We know that just the *presence* of our digital devices next to us distracts us from our family and friends. We feel the need to immediately respond to a text or email. When our phones aren't close by, our anxiety skyrockets, and we experience "phantom vibrations". We intuitively know that our social media use is taking us away from in-person time, and yet because of the dopamine hit that causes addiction, we can't seem to put them down.

When people overuse social media, they suffer from depression, loneliness, narcissism and isolation. Suicide rates have increased in teens, and anxiety levels have skyrocketed. There is a clear link between these statistics and excessive digital device usage.

Sleep, strongly affected by anxiety, is being deprived due to excessive device usage. This basic human need is regularly being compromised by our incessant use of digital devices. We have become a distracted, addicted, and unfocused society—directly correlated to our digital device usage.

I noticed the obsession with digital devices when it began to enter my therapy room daily in sessions with my clients. Stories like Jenny and Tom had become commonplace. I soon became curious about the effects of texting and swiping on our relationships as well as on individuals.

This inspired me to do my doctoral project on technology and dive into how our tech influences us relationally. Professionally, I work with companies to help them manage technology distractions. This improves production and establishes real connections in the workplace. I also have a private practice as a marriage and family therapist, where I work with families to establish authentic connections which last a lifetime. I help them learn to disconnect from tech so they can reconnect with self, family and careers.

As I began to face the challenge of having teenage kids myself who would rather text me than call me, I was even more concerned about technology and its impact on our family. I knew I needed to find creative ways to stay connected with my kids. My family could have been a case study for addiction related to technology. This inspired me to learn and gain the tools to fight for my family. Along the way, I developed the skills to walk with others struggling as they fight technology-related addictions which have become more and more prevalent in our world.

My own family has been touched and shaken by the dark side of technology, and because of this, I am able to be sensitive and empathetic to the pain that can come from screens. I've seen how using them excessively and improperly can steal our souls and leave us feeling desolate and empty.

As a society, we are in the fight of our lives for our minds and the minds of our children. The daily barrage of technology often overwhelms our vulnerabilities and decreases our capabilities to live healthy, clean lives. Tristan Harris says that as we upgrade the technology in our lives, we downgrade our humanity. No one is coming to rescue us from the harmful effects of tech; we must save ourselves. We must look at technology from a different angle beyond its convenience and instead at the ways it affects our minds and lives.

We must examine our mindset around technology and adjust it based on current research—not media hype over the latest must-have apps. It is time to become educated about the effects of too much technology.

We are wired to connect. It's who we are biologically. At the moment of conception, we begin to thrive because of our connection with our source of nurturance, our mothers. After birth, we grow spiritually, emotionally, and intellectually when we make healthy connections.

This is not speculation; it is neuroscience. The fact that we are wired for connection makes it very unhealthy when we enter *dis*connect. Technology contributes greatly to our disconnect as it offers us only a "fake connection." Our brain doesn't know the difference between real and imagined connection; it may believe we are connected through tech, but connecting through tech isn't the same as real, in-person connection.

Just because we are online together doesn't mean we are really connected and engaged. In fact, quite the opposite is true. A like on Facebook is not the same as love in real life. When we go to dinner with our kids and everyone has their head down and fingers scrolling—the new human position and condition—it begs the question: why even be out together? Eating a meal together once a day contributes to our overall well-being but a meal with digital devices present is a poor substitution for conversation and connection.

My message about technology is one of hope.

I believe that we can all do better—both the tech creators and the users. This book will teach you a collection of best practices that have been developed to decrease excessive device use and give us back our lives. They have been tested in my office and in the world with real people who have experienced real, human connections.

There is a movement in our world that understands the need to monitor and manage digital device usage. It is growing stronger each day and holding the creators of tech accountable. It's also helping parents and individuals establish healthier relationships with technology in their homes.

As you read my book you will become aware of the ways that social media has changed many aspects of society. You'll learn what you can do to break up with your digital devices and become engaged and connected with yourself and others in real time with real-life relationships. This is authentic to how you are wired as a human being.

I'd like to invite you to become knowledgeable about the ways your mind and the minds of those you love are hijacked by tech. I want to teach you what you can do to prevent this. I want to show you how you will benefit from less tech time. You'll discover a plan to reconnect with yourself that will help you enjoy greater well-being, intuition, and increased confidence. You'll build a community that empowers you to live a life you love. You'll improve focus and be a better communicator. Your life can be richer and more enjoyable, IF you are tuned into life and turned off of tech.

When we are better connected and engaged, we feel heard and seen. We also see and hear others, and that builds connection. The stronger our in-person connections become, the more balanced our mental health and emotional well-being become.

Conscious connection is everything.

Our world's need for love and human connection is never more evident than in the story on ABC news of a football coach at Parkrose High School in Oregon, who disarmed a student with a shotgun and embraced him for over a minute.

Video footage showed the emotional hug the two shared. It's a reminder of the gift of compassion and the importance of human connection. Connection is what our hurting and chaotic world needs. We can't do this when our hands are full of our digital devices and our minds are distracted. It only happens when we put our devices down and become connected and engaged.

Your time is limited, and how you spend it matters. Spend it looking into your lover's eyes, having lunch with your best friend, and taking trips with your family. Detox from digital device usage, and get to know one another in real life.

I hope this book helps you disrupt your unhealthy tech habits and take action to enjoy connection and happiness in real time.

When we know better, we can do better. Let's do better.

CHAPTER ONE

Relational Connection

Some people cross your path and change your whole direction.

—UNKNOWN

TOM AND JANE came to me because they have a problem.

Every night, when Tom gets home from work, they become creatures of habit. As they prepare their dinner in the kitchen, they watch TV, listening to the horrors of the day from the newscasters. Then they sit down and eat together at the kitchen counter, both with their iPads in hand. They take a bite and scroll. Take a bite and scroll.

Very few words pass between the two during this time. After dinner, they go to the living room, turn on the TV, and watch it together for a couple of hours. Finally, they climb in bed for the night with their Kindles to read for a while before falling asleep.

Tom and Jane complain that they feel disconnected. They know they love each other, but they simply can't understand why they don't feel closer, why they aren't better connected.

As I heard their story, the words of Adele's song popped into my head: *Hello, from the other side* could describe their relational dynamic.

During the day, they pass texts back and forth. They talk about what they will have for dinner and who paid the gas bill. When they send a letter to one of their kids, they copy the other via email. They update their extended family with an email letter once a month.

In their minds, this should make them more connected, yet they feel so distant from each other.

This disconnected experience Tom and Jane describe is not new to me.

I hear this almost daily in my practice. Families are more plugged in than ever before due to technology, and yet they're less connected. Consequently, they feel far apart. There is a chasm; they *feel* it, yet they can't seem to understand how to *fill* it. They come to me to help them reconnect and fill the void.

Tom and Jane are not the only people on our planet, with its new and faster 5G technology, who feel lost trying to find their way to better connection. Most of us walk around with a mini-computer in our pocket—our smartphone. Yet how smart are we if we are so involved with this device that we lose touch with the world around us?

We know more people around the world have cell phones than have ever had landlines. There are almost as many cell phones in operation as there are people on the earth. It's an epidemic.

How is this affecting us individually and in our relationships? You may not have the answer, but you can certainly understand the question.

Considering that more people have cell phones in our world than have flushing toilets, (yowsa!), the prospect of its effects is scary. Is this a silent disaster lurking to disconnect our world even amid the most rapid technological connection ever? Are we connected in ways that are ruining our relationships, health, and finances? We are currently the most obese and addicted society in history. Is our disconnect a contributor to this unfortunate statistic?

Consider your own life as an example. What is the first thing you pick up when you awake in the morning? A large part of the world's population would say their smartphone, iPad, computer, or TV remote. For many, their device is the last thing they look at before going to sleep and the first thing they look at when they wake up.

For twenty-five years, I have studied human behavior and heard hundreds, if not thousands, of stories about the disconnect people feel. I've seen it firsthand from my colleagues, professors, and clients and even in my own personal life. I've discovered that we have many ways of disconnecting from ourselves and our loved ones; some we are conscious of, and some we are not.

This book will teach you some helpful ways to connect— ways to bridge the divide that affects each of us relationally and individually.

Of course, making positive changes involves sacrifice. Sacrifice is work, and work can be uncomfortable. We have misled the next generation to believe that it is not normal to be uncomfortable. Sometimes we are uncomfortable, and there is not one darn thing we can do except sit with it. Everything can't be fixed. Some things just *are*. In these times, circumstances don't need to be fixed, we do.

Freedom is not free. None of the wonderful things we enjoy in life have come without sacrifice. If I want to be healthy, I must make changes in my daily eating and health habits. My desire for health must make me uncomfortable when I want to binge Netflix all day from the comfort of my sofa. If I want to enjoy a healthy life, I need to stand up, get moving, and make healthier choices.

Personally, I often have to disengage from my tech and go for a walk. No pain, no gain. It's just the way it works, kiddos. Life ain't easy, and neither is freedom. We are not entitled to anything worthwhile; we must earn it. We operate so much better when we work to gain that which we desire. This is what being healthy looks like. It's been my observation that the unhappiest people in the world are free riders with no purpose and no motivation to be connected and engaged.

There is usually an internal force that we have to challenge in order to disconnect from our tech and engage with the real world. This is the challenge we face when overcoming any habit.

When I engage with others face-to-face, I learn they may think differently than I. That's ok. Everyone doesn't have to be just like me. How boring would that be? Why, oh why, do we get on social media and argue with others because they think differently than we do? Why can we not simply appreciate one another's unique differences?

Differences are what makes our world go 'round. Why do so many people have an—*I am right, therefore you are wrong. Now go away*—mentality? We must learn to tolerate other views and accept others as people.

Relational connection means engaging and connecting with people in loving and healthy ways. Maturity means giving up the need to make others like you in order to live peacefully in community.

No matter how old we are, we all need to grow up sometimes. Growth is challenging. We can connect the unconnected creatively and respectfully. When we successfully connect one with the other, the connection must be mutual and travel both ways. Connection isn't one-sided. There is no connection without mutuality.

You have to be willing to be a little uncomfortable for us to live together peacefully. I may need to accept that you are you and I am me, and we don't have to think the same way about everything. I have to realize that I want the freedom to express my unique thoughts but that you may not agree with my views, and I may not agree with yours. This reality means I may have to accept that your views are different than mine without being offended and enraged. You be you, and let me be me, and then let's meet in the middle in love and harmony.

Life is all about connections, and it is so much better in community. A good connection is like winning the lottery. We must bask in our good fortune when we have connected well and authentically.

You bring to the table a set of values colored by your own past experiences. So do I. That is what makes us each unique. We must honor the uniqueness of others while staying true to ourselves.

Dissatisfaction will look different for each of us. The key is knowing that it will come into our lives along with discomfort. That is part of life. I can't change you to fix my dissatisfaction. It's not happening. Nor should it.

Discomfort is where growth begins. It is not a problem to rush to fix, but instead a gift to be embraced. Our ability to reason grows through problems. Our strength is challenged and grows, as well.

We have become a society that panics over differences and problems as if it's realistic to expect neither to be present in our daily lives. As Scott Peck said, "Life is no bowl of cherries." Who told you you should be happy all of the time? Expectations are everything. If you set yourself up to expect an always-happy life, you will be sadly disappointed. What will you do with that disappointment? Will you spin it into anxiety?

As we grow and mature, we will realize that the freedoms we have will cause each of us to learn about compromise and acceptance of the views and lifestyles that are different from our own. That is the epitome of freedom. You can't have freedom without differences. Maturity is recognizing this.

Democracy is about everyone getting a vote. When we can't honor our neighbors' votes, and we instead try to make them think like us, then we are not honoring democracy. Democracy should afford us all to live peacefully together. The current divide in our world is about me, and my expecting you to be like me so I won't feel uncomfortable.

Until I realize that completeness looks a little like your way, and a little like mine, then there will be no harmony and no authentic relational connection.

CONNECT & ENGAGE: *Evaluate your level of relational connection with the people you value most. Where do you need to invest more energy, and what might you cut out to make that happen?*

CHAPTER TWO

Sleep Smart

*Each night when I go to sleep, I die, and when
I awake the next morning I am reborn.*

—MAHATMA GANDHI

MARY CAME TO me and shared that she is tired all day long. She constantly contemplates going home and climbing between the sheets for a restful night's sleep. But when she does get to bed, sleep escapes her.

She tosses and turns, thinking of the unfinished tasks from the day. She reviews the long to-do list she will have to tackle first thing in the morning. When her feet hit the floor, she is exhausted before she even climbs out of bed. She knows she will spend another day disconnected and distracted. She sighs as she throws the covers back.

This is the story of her life lately.

At night, when we are sleeping, something powerful and important is happening. This is when our brain heals from the anxiety of the day. So, you can imagine what happens when we don't get a good night's sleep. Stress and anxiety build and can snowball into the next day, and the next, and the next...

When our sleep environment is not adequate to facilitate a restful night's sleep, then we suffer the following day. As William Shakespeare said, sleep is the "balm of hurt minds," a chief nourishing element in life's feast.

We desperately need quality sleep, and technology often interferes. Many of us view sleep as something that gets in the way of our internet surfing and TV watching. We don't realize the peace we can experience as a benefit of a good night's sleep.

Researchers are learning that without a good night's sleep, our health begins to be compromised, and we struggle in our day-to-day lives. Sleep disorder diagnoses are on the rise. With over 71% of us sleeping with our phones by the bed, it has become a large factor in sleep disorders. Anxiety and depression are symptoms of sleep disorders, and anxiety is the number one mental illness in our world today.

Is it a coincidence that we all have smartphones in use constantly, *and* we also have higher than ever anxiety issues? I highly doubt it.

Do you need a nap every afternoon, wake up tired in the morning, or fall asleep while watching TV? Then quite possibly you need to curb your device usage at bedtime.

See if you can relate to a few other symptoms that manifest when tech interrupts your sleep:

» Irritability
» Trouble focusing during the day
» Needing alcohol to fall asleep
» Difficulty falling asleep or staying asleep
» Consuming multiple energy drinks
  during the day to stay awake
» Feeling that your time is not productive

If any of the above describe you, then you are likely not getting proper sleep and should consider less screen time to support your sleep health.

Research supports that if you spend more than *one hour each day* on your device, you will be more likely to exhibit symptoms of depression, which is the twin of anxiety. Too much device usage an hour before bedtime can cause brain stimulation, which interferes with REM sleep. The best way to remedy this—set your bedtime, and don't watch your screen for the hour prior to sleep. Don't take screens into the bed with you or watch them from the bed. Simply use your bed for sleep, to enjoy the peace of a good night's rest.

Women need approximately seven hours of sleep, and men need eight hours of sleep. This ensures you move through each of the REM states to provide peaceful rest. All sleep is not equal; we need *high-quality* sleep to keep us at our best. Women suffer more side effects with lost sleep than do men. For example, women who sleep fewer than 5 hours each night are *45%* more likely to have heart issues!

Losing sleep can be a slippery slope that leads to other problems. We suffer brain fog and decreased creativity and productivity. Of course, this impacts our careers and private lives greatly. Even our appetites are affected by poor sleep. We eat more and sit on the sofa watching TV or surfing the internet, a major cause of our increased obesity as a culture.

Our moods are affected because our bodies secrete less of our happy hormones. Serotonin and dopamine levels fall when we are sleep deprived. These brain chemicals help us feel secure and stable. Without them, we crave sugar, and our anxiety skyrockets.

Down the slippery slope we go.

Too much tech can be directly linked to a sedentary lifestyle. Your appetite increases with each unhealthy sleep cycle. Your immune system suffers when you are not sleeping well. You need a healthy immune system to fight off disease. Research indicates that if we get less than six hours of sleep, we raise our risk of viral infection by 50 percent.

Lack of sleep raises levels of inflammation in the body, which increases the risk of heart disease, cancer, and premature aging. Poor sleep can increase blood pressure, which also puts us at risk for stroke. We know that 50 to 70 million people in the US suffer from alcoholism, anxiety, depression, and mood swings, which also raises the risk of stroke. This is because the body does not get the chance to restore organ function and chemical imbalance. This failure contributes to behavioral and mood problems, all of which our culture is experiencing in abundance.

Think of your brain as a computer. At night when you are sleeping, your brain is repairing circuitry issues and rearranging your experiences from the day, much like a computer manages data. As humans, we are designed to get that needed sleep every single night. It's part of our physiological makeup. Yet seventy percent of us are going without proper sleep. With the patients I've seen—and I have seen many—this is directly connected with the technology they interact with throughout their day.

Our systems become buried with sensory overload, and the place this becomes most obvious is in our health and sleep habits. We reach for a cup of java in the afternoon to help us get through the day, but because it stays in our system in excess of 6 hours, we are not sleepy at bedtime. We turn on the TV to get sleepy, but that is like adding fuel to the fire. The screen stimulates our brain and interrupts our sleep throughout the night. We wake up grumpy and tired.

Small changes can bring big rewards.

The next time you can't sleep, try reading a book. Or get a journal and spend some time writing down your thoughts and reflecting on things with gratitude. Eliminate screens in bed for at least one hour before bedtime. You'll begin to see improvements in your relationships, productivity, and creativity. With a good night's sleep, you can manage most anything life throws your way.

CONNECT & ENGAGE: *Think about your routine right before bedtime. What do you do that sets you up for sleep success? What do you do that might hinder your sleep? What small change can you implement tonight?*

CHAPTER THREE

Focus

What you stay focused on will grow.

—ROY T. BENNET

ROXIE, OUR SWEET family dog of fourteen years, was so very sick at the end of her life that the vet had placed her on many different medicines for her congestive heart failure. At the time, I was knee-deep in research and working around the clock to complete my Doctoral project. There were so many days that I was distracted and stressed that I had very little time to focus on anything except my project.

One day, I set my vitamins out and then placed Roxie's meds beside them on the kitchen counter. The phone rang, I answered, and had a conversation. When I hung up, I went back to give Roxie her meds. I looked down and realized that I had taken my handful of what I thought were my vitamins while on the phone. But now I could see my vitamins still sitting there while Roxie's meds were gone.

Oops.

It appeared that in my multitasking and distraction when the phone rang, I had taken Roxie's meds. One was for water retention, one was for heart issues, and one was for anxiety. Oh my, what to do? Thank goodness the vet quickly answered my call and assured me that I would be fine.

It is really hard to focus when you are tired. That morning, I had a good excuse for my tiredness, but I've also been tired and unfocused because I have spent too much time on tech. Too much internet surfing affects our sleep negatively, but it also affects our attention span and our ability to focus. When device usage becomes a compulsion, it can contribute to ADHD and other focus issues.

Think about it.

How often do you notice yourself texting more than you wanted to, or planning to look at social media for just fifteen minutes—which then turns into hours. Before you know it, the afternoon is wasted. Has your device devolved from something that should bring ease to your life to something that causes anxiety and angst? Maybe it has, but you haven't even noticed it yet.

Each time we look away from something we are doing and check our digital devices, it takes us approximately *twenty minutes* to refocus. This constant checking and rechecking of our devices is destroying our attention span. Imagine how we could better use that twenty minutes. I encourage you to designate a time each day to check your device, and don't look at it otherwise. You will build your focus muscle and use your time more wisely.

Students especially are negatively affected by compulsive device usage. Their grades often suffer as a result. Too much device usage can alter their ability to focus and shorten their attention span. Our average attention span used to be twenty minutes, and now it is down to eight seconds. Research has discovered that our attention spans are shorter because of our increased exposure to technology. Nobel Prize-winning economist Herbert Simon said, "A wealth of information creates a poverty of attention."

Our kids are particularly sensitive to the impact of screen technology on their attention spans. Childhood is a crucial time for brain development, and too much screen exposure interrupts healthy brain development in some children. In the psychology field, we've seen an increase in ADHD diagnoses and an increase in Ritalin and Adderall prescriptions over the past decade. There is a correlation between this and children's overexposure to screens.

Human brains have evolved to adapt to their environment, and when children don't have to exercise their brain, or they do it in a different way, it changes their development experience. Neuroscientists believe that changing the way kids learn affects the brain in many ways.

Think about this. Just a couple of decades ago, a child writing a research paper in middle school used books and manually explored for the information. Now, a Google search is all that is needed to gather research. As a result, different areas of the brain are stimulated through this new

Google experience, and kids miss out on the search-and-find element of research and its benefits.

Children's brains are malleable and function differently now due to screen technology. Their brains are substantially formed by the experiences in their daily lives. This growing attachment to technology is altering the way we think and feel, and it shortcuts brain power. Teachers of school-aged children are reporting shorter attention spans in their students due to screen-based activities in lieu of reading conventionally.

The Associated Press has requested that its writers develop shorter stories due to the fact that they have noticed readers are more likely to read the abbreviated versions, as opposed to longer stories which require greater attention spans.

We can no longer sit still long enough to go deep.

The human brain is under attack from the modern world of technology. This crisis can change the entire trajectory of your life, as subtly it is changing the world around us. It affects who we are, what we do, and how we do it. It shapes how we act with one another and how we connect and engage with ourselves, others, and the world around us.

If your phone is your most intense relationship, it might be time to break up.

Electronic gadgets are turning us into zombies. Our brains are like plastic; the way we use them determines their shape. The brain changes in response to external stimuli. Technology has an effect on the microcellular structure as well as the very detailed biochemistry of our brains. And that, of course, affects our focus.

We know that our brain has an uncanny power of imagination when it is focused. In many ways, technology is circumventing that focus and interrupting creativity.

Focus is critical to success in any area. If we begin our day with a plan of what we want to achieve, and remain focused on the goal, we can move forward. Having and pursuing a goal every day will bring the change you wish for into your life.

Life can be difficult. Sometimes you can't anticipate the way the day is going to go. Anxiety rushes in, which causes distress to grow. If you can remain focused on your one goal and continue to push towards it, you will find peace from this accomplishment. I have become a creature of habit and am mindful to incorporate healthy habits into my day. I begin by making my bed as soon as I wake up. This one simple task sets me up for success and begins my day positively. The rest of the day may not go so smoothly, but I have started with a positive mindset that helps me handle whatever the day brings.

Each day you should identify one thing to focus on that you are confident you can control. Accomplishing it will bring you confidence. Create a sense of structure that provides a foundation for your day to go well. Having a goal and completing it each day brings you satisfaction. You can end your day well and look forward to the next with an uplifted spirit

We know that individuals who have a purpose to work toward fare better in life than those who just float through, with no real reason to get up each day. We all have a gift.

Finding that gift and focusing on using it is uplifting. Sharing our gifts with the world, while others share their gifts with us, is optimal. When you give to self and others by utilizing your strengths, it lifts you up. Focus on the good and move toward being a part of it. This takes intentionality but feels oh-so-good when we accomplish it.

We must realize that once our brains are no longer wired for constant cyber distractions, then they expect focus and look for it. We can teach our brains to learn to enjoy focusing again. Start by scheduling specifically when you will use the internet, then avoid it all other times. I check my email once in the morning and again in the late afternoon. I do the same thing with my phone. I check my messages in the morning, mid-day and in the evening. It's absurd to be on call 24/7. It's unhealthy to be constantly attached to the device that steals your joy and destroys your concentration.

The good news: *you* get to choose. Choose freedom and focus rather than distracted disengagement. It's up to you.

We lose focus on our own growth when we spend hours scrolling through others' lives. This might lead us to abandon ourselves and our own growth. We become so caught up in finding out what our "friends" are doing that we lose track of what *we* should be doing.

Your emotional and personal growth is about you. Sometimes others grow beside you, but sometimes they choose not to go on your journey. We have to accept that others may not join us on our journey. We must remain focused on ourselves and our personal growth.

Just because someone else is not joining us on our journey does not mean we are leaving them or don't care about them. Our integration is our work, and, although we invite others, it is ours alone to remain focused on. Our growth is about focusing on reclaiming the "us" we have lost along the road of life. It is about owning our intimate and deep understanding of self. To do this we have to acknowledge our conditioning and programming, and then uncover the things in our lives which need to be unpacked for our journey into self. We must come to understand all parts of ourselves and integrate the trauma, wounding, and pain that tries to be in charge of our lives at times.

When we remain focused on personal growth, we operate with more understanding of how our actions and choices impact the world around us. We learn to expand rather than contract. We grow rather than becoming small. We become connected and engaged with both self and others. We come into ourselves and find it to be home for us.

Growth is not about rejection of self, but rather acceptance of who we find ourselves to be. When we leave behind the beliefs which limit us and leave us stuck, we sometimes become aware of how the pain of others has changed our path. We no longer focus on patterns that don't serve us; we leave those behind. We focus on new and better ways to navigate our process of self-growth. The things in our lives which we determine are not serving us well get left behind as we focus on the new way that we know to be better for us.

Sometimes our thoughts are the first things we need to leave behind. For many, their thoughts are primarily unhealthy self-talk. When you talk to yourself negatively—no big surprise here—it affects your brain in a negative way. It can change your brain into a state of anxiety. Saying negative things to yourself also affects your body physiologically.

Those who live their best lives are positive in the present. They attract the abundance through their thoughts. Anytime you use the words *I am*, you are unconsciously programming your mind. For example, saying *I am well* will help you feel better. Saying *I am healthy* is a choice and will bring health into your life. Your beliefs are a choice. Internal good feelings are created by the beliefs your focus on. What do you focus on, and what does it say about your future?

When you want your life to be better, focus on changing your beliefs, not your behavior. We don't come into the world with a manual about how to operate our brains. One of the fundamental rules for life is to know what you want and then focus on it. The motive for your action must be a reason beyond yourself.

For example, I have learned so much in my studies and through obtaining my Doctorate. It's helped me become an expert on human behavior. This is not a path to get rich, but it is a path of joy for me. I find myself driven to share the knowledge I have gained with others. I want people to have the tools they need to have a better life. I want to help you achieve the life you wish for but don't realize is possible. It begins by focusing on your thoughts.

In the game of life, we have to decide to what level we wish to take our lives. If you have dreams but you are unfocused, be prepared to struggle and pay a great price. Unfocus limits your potential.

By focus, you change your potential. Start small and give yourself one command to build your discipline muscle. Decide to do it now. Give yourself a command, and then follow through. This creates a neural path for success. You are training your brain to be in control when you focus on your goals. When you do this and see your success, you begin to believe in you. You leave a positive imprint on your confidence. Everything you do either prepares you to move toward your goals, or it moves you further from them. Teach your brain to make the positive patterns automatic.

At the end of the day, do you feel like you have been needled all day long by the constant beeping of your phone and its myriad notifications? Emails. Ding. Snapchat. Ding. Messenger. Ding. Instagram, texts, and more. Guess what? You can switch it off, and the world will keep turning. There is nothing so important that it can't wait for you to pursue your goals and dreams.

Those of us who already suffer with the "I'm not enough" syndrome feel obliged to be superwomen and supermen. We think we need to be on call every single minute. What if someone needs us? What if we need to save the world? You can see how ludicrous this sounds, but if you're honest, you realize you have thought this way before. Why do we need to be digitally connected 24/7? It's not healthy. Many people think this "always on" mode will help them get ahead, but if you are able to be more focused and peaceful, it will move you past the competition. They are chained to their phones, while you are head down and getting things done.

Find peace. Dial off. Unplug. Be present. Show up for yourself this way.

An important aspect of focus is discipline. We must be willing to be disciplined to reach our goals. You will not find a successful person who *wandered* into success. Discipline looks like this: staying up an hour later than your partner to write a book. Missing a really fun party to get your daily page quota completed. Passing on the cookies because you have a healthy body in mind.

Discipline almost always requires giving up something to get something. It's delayed gratification—that thing we least like, and society trains us to detest. Our cultural expectation is that I want it, and I want it now. Guess what? It ain't happening. Life doesn't roll that way, and success surely doesn't roll that way. It may be half a decade before you get that thing you dream of. Will you quit? That is how discipline, reality, and success collide. That is why it's critical to set yourself up with realistic expectations. Be willing to do whatever it takes to drive and focus and get that thing you want.

Nothing stays the same in life. Focus on that thing you wish for, and hold steadfast to that visualization. Don't rest, but instead stay in motion. Proper motion means growth. But growing without focus is not likely. Build something in your life through focus. Will yourself to focus, and be disciplined to change in the way you want. Dissatisfaction is often the very best motivator for positive change.

What change do you want to see in your life? Are you disciplined enough to get there?

When you get dissatisfied enough with where you are in life, you become motivated to focus on the change you want to make. You get serious and do what it takes to get there. When couples come in to work with me, I can tell pretty quickly which couples will do well in therapy and which will not. I simply measure their dissatisfaction in the relationship and their motivation to change it. If they are both serious about making a change, there's a good chance I can give them the tools to make their marriage better.

Setting a goal is important; staying focused is a must. Focus helps you examine how you are living and what you are doing with your potential. Comfort is a deterrent to growth. Focus on the places in your life where you are most uncomfortable, then walk through them rather than around them. Use your imagination to stay focused. Visualize that you are the person you wish to be and remain focused on that vision.

Progress won't be easy. In fact, it may be really hard. You don't have to know every step to get to the place you are focusing on; just do the next best thing to get there. Believe and you will receive. Focus on that thing you believe in with no limitations.

The only prerequisite to your success is this: do you want it, and can you remain focused long enough to get it? Nothing else matters. You don't have to know how you will get there. The opportunities will come, and the right path will appear *if you remain focused.*

Be willing to explore all possibilities toward your goals. All men dream, but we don't all dream equally. Successful people focus on their dream and prioritize it. They don't sit and wait for it to fall into their laps.

The most important part of your journey is today. Concentrating and focusing makes the intangible become tangible. The more you focus, the more tangible your dreams become, and the more connected and engaged you become to all the possibilities before you *right now.*

Through focus, you can create a mental environment that brings you peace and clarity as you move towards your dream. Be mindful. It is the very essence of focus. Build your focus muscle by remaining true to your dream. Be disciplined to move toward it. Once you are fully engaged in this way, focus becomes your path to success.

CONNECT & ENGAGE: *Perform a focus evaluation. What things constantly distract you, and how can you minimize their presence in your life? Where do you need to clarify your goal or dream so you can begin to focus on it and visualize success?*

CHAPTER FOUR

Choices

We are our choices.

—JEAN-PAUL SARTRE

JANE WAS IN midlife and came to me to process where she was in her life and how she got there. She wanted clarity on how certain aspects of her personality were formed and what contributed to them. She wanted to understand what led her to make the choices she had made thus far—the good and the bad.

Jane was like all of us; she had been presented with several possibilities in particular situations and then had to choose the one that was the best fit. Sometimes it worked out well and other times, well, not so much. Regardless, Jane knew that understanding the *why* behind her choices would be significant for her future choices and growth.

In my practice, I see many who come in due to anxiety. As we unpack things in their life, we realize their anxiety is quite often tied to a choice. Life is all about choices, and where we are in our lives is a culmination of the choices we have made.

Some become so afraid to make choices that they have "uncertainty intolerance." They *need* to know that their choice will work out well before they make it. Of course, the final outcome of a decision is impossible to predict. We have no control over certainty, but we do have control over how we handle uncertainty.

If we are fortunate, our parents began early in life allowing us to choose. This helps us build a muscle of tolerance for when we don't make a great choice and things don't go well. This is a natural part of making choices. We must learn how to react to our poor choices and right ourselves when things don't go as planned or how we had hoped.

Perspective is important. Seeing uncertainty as part of life and knowing things will sometimes turn out poorly is a mark of maturity. How we handle that is critical.

Why is this chapter on choices in a book about disconnecting from digital devices? It is important to free our minds of outside chatter and be connected and engaged with self when making important choices. Sometimes, the first place people go to get help making choices is online. Then that choice is not your choice.

It is hard to make a wise choice when you have so much external chatter. There is great value in getting quiet and sitting with yourself to find the answer you are searching for. External validation about a choice (where you are internally conflicted) can only lead to frustration and confusion. This builds anxiety. You do you. Find your answers inside yourself, and feel the freedom in making your own choice. Whether it works out or not, it was yours.

At the end of the day, do our choices bring us into love? Many are perishing every day and hour, because they have no love in their lives. We each have a universal desire to love and to be loved. According to Maslow's Hierarchy, it is one of our basic needs. Being connected with ourselves so we are aware of the amount of love in our lives and how we fill the void is imperative. There are ways that we pursue love which leave us feeling lonelier than ever. From our family of origin, we learn maladaptive ways to find love. But we learn good ways to find it, too. We have to search our past to understand how it is present in our lives and if it is leading us to love or a void.

How did you learn to be loved? Was it in a healthy way that you can use for a lifetime to make authentic love connections? The way we love is important to how we experience life. It is a measure of the love and joy we have, as well as the peace.

To understand how we learned about love, ask yourself these helpful questions:

» How did you know your parents loved you?
» How did you need to behave to gain their love?
» Who did you have to be to be loved by your parents?
» How do you express love to those you love?
» Do you find it easier to give or receive love?

If you can understand the answers to these questions, you can learn to build bridges and drop the walls that keep you from experiencing the love you wish for and deserve. You are the gatekeeper for the love you have in your life. Be curious and grow.

Once you have decided to open your heart to give and receive love you will be amazed at how this begins to unfold for you. When you let go of limiting beliefs that keep you from the love connections you wish for, you will find the love you need and hope for. Are you willing to admit that you need love as well as you need to give love?

Do you tell yourself stories like:

» No one will ever love me.
» I don't deserve to be loved.
» I will always be single.

These stories are defining your life and keeping you from giving and receiving love. Why do you carry such thoughts, and where did you learn them? These thoughts don't just pop into your head from nowhere. They have roots, and once you understand those roots, you will be able to overcome these limiting beliefs and enjoy the life you are yearning to live. You must learn how your past is present in your thoughts and behaviors and how it is holding you back from living your best life.

Our choices determine our self-care. When we are well-connected to others and self, we put our self-care at the top of our to-do list. We also make time for healthy relationships which enrich our lives. We build good connections. We recognize what is on our list of things to do that is significant and what can be deleted. We know passing some responsibilities to others frees us and empowers them.

You don't win by being the most exhausted person in the room, but instead by being the person with self-care intact. You are not defined by the length of your to-do list. Learn to prioritize and know the difference between real priorities and busyness. Wise choices help us pace ourselves and should include fun in our daily care and schedule.

We can make choices that eliminate our stress and help us to be better connected and engaged.

The circumstances in your life often put you in a position to choose how you respond. Reacting is never a healthy choice. We can allow circumstances to bring resentment and anger, or we can let them make us kinder and more human. When we are connected to self, it is a tool to assist us in dealing with the difficulties of life. Choosing to be connected and engaged ties you to others during hard times, and there you find your strength to move forward. Make friends with fear, face the darkness with curiosity, and choose carefully how you will respond.

The choices we face and the internal warfare which accompanies them are a reminder that we have low tolerance for being uncomfortable. This influences the choices we make. Sitting with our unease and making friends with it disempowers the ghoulies.

When we choose to stop running, but instead turn and face that which we fear, it stops chasing us. Move closer to your emotions and thoughts, and you'll know the best choice for the ending you want. Make the right choices for you, and you change the ending to your story.

CONNECT & ENGAGE: *Think about the way you make choices. Do you have a clear understanding why you choose the things you do? Where do these choices come from?*

CHAPTER FIVE

Communication

I think for any relationship to be successful, there needs to be loving communication, appreciation and understanding.

—MIRANDA KERR

WHEN MY DAUGHTER was on her way over for dinner one evening, I asked her to stop at the store and pick up a few things for me. She jokingly said to make the list short since she didn't have much room in her car. I texted her the list and numbered the items so she would understand that she could go through the express line with under ten items.

About forty-five minutes later I heard her car door slam. She opened up the door and came inside struggling with a lot of shopping bags. She set them on the counter, and I started pulling out the items. She'd picked up one steak (I was cooking for four people), two gallons of milk, three toothbrushes, four heads of lettuce, five chocolate cakes from the bakery, and six gallons of vanilla ice cream.

I looked at the odd mix of items and had to laugh. She turned my numbered list into quantities—a perfect example of the message getting lost in translation. She laughed as she figured out what she had done.

It's good to be loved, but it's profound to be understood.

We all need and wish for communication—really good communication. That kind that builds bridges and fills gaps. Communication is the key to connection.

Communication is everything, but the way we communicate has drastically changed. Less than a decade ago, my kids talked to me. They called me and checked in. I could hear their voices on the phone. Now, they text to update me. In fact, if I get an actual phone call, I panic, because I know something is likely not okay.

That's how they have trained me to think with their texting behaviors. In some ways it is good, because through texts I have written documentation of our conversations and can recall them easily. In other ways, it feels impersonal. Much is lost in not being able to hear their voice over the phone or look into their eyes and notice their facial expressions. Body language is an important part of communication.

Imagine the ways this impacts our relationships. My book *The Effects of Texting on the Marital Relationship* clearly demonstrates the way we impact our romantic relationships through device usage. Not only have screens changed the way we communicate, but also *what* we communicate.

Think about this. If I go to bed at night and my husband is tuned into his device or checking work emails, then on an unconscious level I begin to feel second fiddle and unimportant. Eventually, I begin to bring my phone to bed and together we like and swipe away. We're each in our own world, but they don't touch other than the fact that we are in the same room. It's not really a good way to connect and create longevity in marriage. The joy we wish for in relationships is not possible without authentic, in-person connections.

Our attention spans have dropped from twenty minutes to eight seconds. Since we are so distracted, we are adapting to communicate differently and more quickly. Otherwise, I move onto the next thing that captures my attention—and you aren't it.

Technology has affected our communication in several ways. How we communicate and the quality of our communication are affected the most. The term "phubbing" means the act of being on a device in the presence of another person. Basically, you are ignoring them or giving them half of your attention.

Maybe you've witnessed this yourself. Have you seen a mom on a walk with her kids and she is phubbing? It makes you wonder what those kids are thinking and what they will be like when they grow up. Will they feel heard and listened to, or will they be in my office one day complaining of feeling dismissed? We haven't yet seen the finished product of a generation of kids raised by phubbers. If you are a parent, what message are you communicating to your kids based on how you use technology around them? It matters; it really does.

I recently counseled a young man who was the last child left at home with his parents. All of the other kids had gone off to college. When I asked him how he was enjoying being the only child at home, he said he was lonely. He said at night after dinner there were no more fun family interactions, and so he simply ate then went to his room and closed the door where he gamed until bedtime.

He was seeing me because he felt isolated and disconnected. Can you see why?

Technology is quickly altering how we do business and how we conduct our private lives. It is changing how we relate to one another, and in some cases determines connection and engagement. Many couples who land in my office mention feeling second fiddle to their partner's devices. Connection creates more connection, and we will only experience more real moments with others when we connect mind and soul. This is how we grow our relationships.

And then there is the miscommunication that occurs using devices. For example, texting is rapidly replacing in-person communication, and so much can be misinterpreted or lost over text messages. I advise couples only to use texting for things like where they will meet for dinner and what time to pick the kids up at school.

No important conversations should be communicated via text.

Relationships certainly should not be created over texts.

Perhaps worst of all, we use our digital devices to communicate in a petty way about our values and beliefs. Rarely can you get on social media without being exposed to the value systems and loud opinions of others. Sometimes this problem keeps us in an emotional hole. So many people use their platforms to try to persuade others to think like them. Can we please have a break? Can we agree that others are capable of making their own choices without us bullying them into our way of thinking? We exploit our digital device usage when we utilize it in this manner. It brings division and a never-ending stream of unwanted thoughts from others pushing their agendas.

If we can find a way to govern ourselves, then we can use technology to build character rather than to destroy. We learn so much about ourselves by the projections we spew. Our own woundedness becomes ever apparent in our digital communication. We can know ourselves by how we interact with others online.

I once was in a supervision session observing, and a client continued to badmouth her ex-spouse. Evidently this has been occurring for many months, and the therapist stopped the session and said, "What about you?"

The woman looked aghast and questioned the therapist's statement, as if to say, "Yeah, what about me?" He said, "You are the dummy that married this man you call an idiot, so what does that say about you?"

Yikes. This was really thought provoking. Maybe not the most delicate way to say it, but effective in a blunt sort of way. We spend so much time communicating the shortfalls of others and don't realize that this speaks volumes about us. If we are the person that was attracted to the bad guy, what does that tell us about ourselves? We choose to partner with those who are equal to us emotionally. Look at those you connect with, and you can see your level of health.

Do you see room for growth?

Be intentional about noticing what you spot in others and are sensitive to. This is a red flag for you to look closer at yourself. Why does what they are saying or doing rattle you so much?

An important part of communication is listening. We sometimes think communication is all about talking, but listening is just as important. Some would argue it's *more* important. Our ability to communicate is the biggest single factor that decides the health of our relationships.

Our survival depends on good communication. The information we give to others, and that which we receive, is affected by how we pass it on and how we accept it. To do this effectively, we must listen and be connected and engaged. What we do with this information depends on our own life experiences and how we have learned to interact. The ideas we develop about ourselves early on in life become our springboard for how we manage communication going forward. Being aware of this helps bring clarity to our communication.

Communication is learned. If we find there are ways we communicate which are not helpful for achieving our goals, then we may need to unlearn the unhelpful communication patterns. When we are born, we have a clean slate. We have learned nothing that influences our communication. We have not accumulated habits which influence how we interact with others in our world. We have not learned how we will communicate with ourselves. We learn these habits from those who are our primary caregivers early in life. Quite often, we can see by looking at our primary caregivers the forms of communication that shape us today.

Elements of communication include body language, values, the expectations we have for self and others, and our senses. These are all governed by our brain. Our brain is the most powerful organ in our body, and it holds all of our past experiences which we bring into our present when communicating.

Communication involves a lot of moving pieces and parts. When I talk and you listen attentively, your senses are absorbing what I look like, how I sound, what I smell like, and if you are touching me, what I feel like.

Once you have processed this information, your brain decides what it means to you personally, based on your past experiences and *particularly* those with your primary caregivers. All the while, I am doing the same things. Our pasts are colliding, and that impacts how we manage this communication exchange.

Neither of us really knows what the other is experiencing; we can only guess. For example, you start talking to me and, as you are talking, I begin thinking about how much you remind me of my teacher in 3rd grade. She looked and sounded like you and even smelled like you, too. She was one of my favorite teachers ever. She was always loving and welcoming and encouraged me to grow and learn.

This sets up excellent possibilities for our communication and me being open to what you have to say. Except that on your end of the conversation, you take one whiff of my perfume and it reminds you of your great aunt. She locked you in the closet whenever you visited her, so nothing coming out of my mouth can change that horrid memory and connection for you.

You are done with listening to me, and you stare off into space and think about something else. This all happens in the first few seconds of our interaction. Isn't it amazing to become aware of all that takes place during communication? It explains how we can connect or disconnect so quickly.

Being mindful and noticing our internal dialogue as we are communicating with others is essential. We always have some sort of dialogue going on with ourselves. Until we are able to look at our past and see how it affects our present, then we will not improve our unconscious or conscious communications.

Turning inward and not sharing with others leads to obstacles. This causes isolation and emotional separation. Emotional separation can occur between family members and workmates as well as in friendships. If we can understand the way our communication creates obstacles for us, then we can create better connections with ourselves and others. We must learn to be able to hear what we really are communicating rather than what we *think* we are communicating. When we collectively learn how to hear our messages and voices, we will become open to needed change. This gives our message clarity and helps us be heard.

Sharing with one another how we experience each other is significant for effective communication. Disillusionment is rampant in relationships. This is mostly because our personal expectations are not being met. Quite often they are not met because we do not communicate them properly to others. We don't take time to clearly voice our expectations. It's hard to meet an expectation that you don't know exists.

To successfully communicate, be present. Don't let yesterday's resentments and pain come into today's interactions. Be in the *here and now* when communicating. Being present physically and emotionally fosters greater connection and engagement. Do not let past pain cloud future experiences.

Communication, like any skill, can be practiced and learned. Be consistent in your desire to foster honest communication. Do what you can to listen, pay attention, relate, and communicate, and your relationships will never be the same. They will be much better.

CONNECT & ENGAGE: *Think about a time when what you tried to communicate was incorrectly received. What happened, and where do you think the mix-up occurred? What can you learn from that experience to make future communication more effective?*

CHAPTER SIX

Disconnection

Technology is a good servant but a bad master.

—GRETCHEN RUBIN

WINSTON CHURCHILL SAID he did not listen to what people said, but instead watched what they did. Our behavior is everything, and the message we send others when in their presence is foundational to the relationship and its solidity.

Many people report feeling disconnected. When we start talking, it doesn't take long to see the problem. They can directly connect this to their phone usage and their inability to curtail phone time. Clients complain that they text instead of talk and have very little in-person human interaction anymore. They may send hug emojis, but they really miss giving hugs. It is a lovely thing to meet a person and connect with them on a soul level. We do this best in person.

Do you remember the days when you actually talked to your friends or mates? Connecting builds close bonds. Those days are diminishing; now we engage large parts of our relationships online. We are wired for in-person connection and attachment, so when we don't receive this, we struggle.

One of my favorite theorists, Virginia Satir, said, "We need four hugs each day for survival, eight hugs for maintenance, and twelve hugs for growth."

What's your hug level? How are you doing with this in your life? The devices we are most attached to don't give us these essentials. They may always be there, but they are cold and impersonal.

Dinner time is one of the most important times of day for families. The interaction and connection that occurs is invaluable. It is the cornerstone of turning children into healthy adults. Rarely do you see families at a meal without parents and kids immersed in their devices. They miss wonderful opportunities to make lasting memories.

Parents need to limit device time for themselves and for their kids. No electronics at dinner, and all devices should be given to parents at night before bed. Parents need to ask this question: *Are these devices connecting or disconnecting our family?*

If social media is so social, why do our children isolate for hours to engage? The research is out: it is not healthy, and it is associated with negative mental and physiological outcomes for our kids and for all of us. It is as unhealthy as smoking. If you would not allow your kids to go up to their rooms and light up, then why would you allow them to isolate for hours in their rooms scrolling on their digital devices? Responsible device usage is an important part of parenting now.

Yes, parenting is harder due to digital devices, but when we put a few responsible boundaries in place, we teach our kids and model for them lifelong healthy device usage. Kids need to realize there was life before digital devices came along, and it is not imperative that we be digitally tethered all the time.

Be an inspiration and a positive example for your kids and friends. Be connected and engaged without your digital devices on you 24/7. Model healthy behavior.

The good news is that if we can learn to be better connected, we can set ourselves apart this way in the workplace and in our leisure. Those with focus issues are putting work out there that is less valuable and doing it at a slower place. Learn to be more focused, and you stand out from the others. Focus is key to improving every area of your life—your playtime, your work, your relationships, and your health. Nothing bad comes from being better connected and focused.

We all face tragic times in our lives, and when they come, we need to be connected to get through them. Others can walk with us through hard times and offer us the love and support we need to carry on and push forward. We can't go it alone. We need others, and they need us. We need connection. Finding as many friends in person as possible, and remembering that our well-being depends on this, is significant.

We crave community and connection; we die in isolation.

Being open to new friends in our lives is a beautiful thing, which brings hope. We can keep the old but also embrace the new. New friends bring new ideas, and a new friend with good character can enrich our lives. Long-lasting and deep friendships are great, but so are new and fresh ones. If we don't allow new people into our lives, we are neglecting self and missing out on new and positive experiences. The length of a relationship does not determine its goodness. Be patient. Grow healthy connections by taking the time to tend relationships as they grow in love and patience.

We have no guarantee that the longstanding relationships in our lives will have the good results we desire. When taking a closer look at these relationships, we might realize they are not healthy. For this reason, we can know it is significant for us to open our hearts to others. Find the wonder in new. We can come to see others' character and our own more clearly in new relationships. You do this in existing relationships when you have good boundaries. Shift where needed, and build connection through vulnerability.

As we grow, new people will join our journey. We must be willing to connect with them. A dear companion offers a connection to sanity in a crazy world. Our work on self takes us in many directions. New people in our lives may be more aligned with us on this new journey. Focus on tending the old relationships, but also be open to the new connection opportunities that come your way. You may be beautifully surprised at the new and exciting input others can introduce into your life. But it only starts when you allow that connection.

Healthy families and individuals talk about what is going on in their lives, including the problems. This builds bonds and breaks the disconnect. This is how you find support when it is needed, by talking about your feelings—both good and bad. When we give each other feedback, it connects us and builds trust. Spend time connecting and engaging; it is a fabulous way to show and build affection.

A big part of disconnection is denial. Denial is a defense mechanism we use to disconnect from our feelings. We think by not feeling them that they will disappear. We think those things we need to say will just be forgotten—if we don't say them. Through denial we may be able to avoid discomfort, but not for long. Denial keeps us disconnected from self and others. We won't be healthy and grow without becoming aware and letting denial go.

We think if we push certain thoughts away that we keep them from having power over us, but the opposite is true. Those things we stuff inside will come out sooner or later. They just come out in unhealthy ways if not processed. The things we repress take up hidden space in our hearts and niggle at us. That which we resist persists. The more we run from it, the more it runs after us. It finds a way to push itself into the light, until we have to become connected to it and deny it no more.

When we do the work to look internally and know ourselves, then we are able to feel the feelings and know what we know. What are you keeping silenced? What needs to be seen, acknowledged, and heard? What do you deliberately need to notice?

The truth really does set you free from disconnect.

We resist the things that we often most need to see. We don't acknowledge that which we most need to feel. The things we need to process we avoid because they hurt. We don't realize that going into the painful place is how we begin to heal and grow. Entering into the painful place will require that we change, and change is scary. We will get mad, and we will wrestle with anger, but then we experience the healing as we process the hurtful parts.

It sets us free as it finds its way into the light. Surrender is difficult but sweet in the end. Let go of the idea that someone is going to rescue you. *You* rescue you.

CONNECT & ENGAGE: *Think about your connection with others. Where do you need to connect more? Where might you need to disconnect?*

CHAPTER SEVEN

Pain

Turn your wounds into wisdom.

—Oprah Winfrey

To do my job as a therapist, I have to be able to leave myself and my world. I have to enter into the world of my clients. This helps me hear them and know them. Only when I can bridge the gap between me and them can I begin to know their pain. This is critical, because I cannot help them until I know their pain. You see, they know when I am there with them authentically, connected and engaged. They encounter me, and I encounter them. We are joined and connected, and the space between us is now sacred. It is where we meet. When I can't understand their space and am unable to enter it, they feel judged, and we lose connection.

Every day on social media we pollute the space there. We bring our pain to the world, and we spew it there for all to see. Most people take no responsibility for this behavior. They are so focused on the pain that they don't see or care how they have treated other people. They strike back, and then the hate builds and grows. It spreads in a nanosecond throughout our world.

What if we changed our way of interacting?

What if instead we brought our open heart and generosity of spirit? What if we said, "I hear you, and I am open to learning about your world." When the other side is taken out of a place of defense, the two limbic systems connect, and our emotions begin to bridge the gap.

Our central nervous system actually begins to calm down. Our brain is the only organ in our body which behaves due to external circumstances. Our mirror neurons begin to form new neural pathways based on our interactions with others. What we see, we often copy. This makes us more relationally intelligent. So what if our online experience were to change from one of hate and degradation to one of positive connections and the meeting of two souls?

Unfortunately, we have adapted to this life of online hate and pain. It's become the new normal. It is easier to hurl hate at those different from us than it is to try and see them and hear them. One man I was counseling grew in his understanding when he said of his experience with those he held contempt for, "I used to be different, and now I am the same." He realized that in joining with another and trying to understand their pain, he actually grew as a person. He learned compassion and empathy and began to enjoy the connection and engagement around others and their differences.

I've met with a client weekly for a couple of years. We have walked through so much together on her journey. She's moved away from abuse and mistreatment into self-love and confidence. At the end of one particular session, after she had an "aha moment," she paused and looked at me before speaking. Then she told me she loved me.

Her heart was so full that she was finally able to feel safe. After years of living with a man so horrible that she could only wish for survival, she had gained back the capacity to love. As she shared this, I knew that she wasn't asking me to love her back. She simply wanted to revel in the fact that she had crossed the bridge from a decade of pain into a place of healing. She was connecting once again.

As we sat together, I simply let her love fill the space. She had worked so very long to find the courage to finally connect with another. To be able to drop her defenses and find safety and connection was her liberation from the hate and unkindness she had encountered for so long. She was able to know that even from that hate, she had grown. The thing that was meant to break her had actually given her strength and courage and insight. She understood others and herself. Connection had made her whole.

Every day we have the choice to give others the gift of kindness. We choose how to fill the space between us. The poet Rumi says, "Beyond right thinking and beyond wrong thinking there is a field. I will meet you there." If we can go online and meet others in this field where we leave our stuff behind, we can find peace as we offer kindness. We get to choose.

When our feelings are hurt, we experience this in the same area of our brain where physical pain is registered. For example, if we are rejected or not included, our brain does not know the difference in physical or emotional pain. It just hurts. It is interesting that we are wired to recognize social rejection as an emergency that we need to address ASAP. This research helps us understand how important connection is to our being. Naomi Eisenberger writes, "There is something about exclusion from others that is perceived as being as harmful to our survival as something that can physically hurt us, and our body physically knows this."

We need to be connected to one another. When we are not connected, it affects our health negatively—and even our lifespan. We can die from disconnection. The need for physical connection is innate. Online connection is not physical. To be connected, we must be able to let go of all which emotionally holds us back from connection. You have to manage your fears so that you can seek others and bond. You have to let go of insecurities and overcome the lack of confidence that holds you back from engaging with others.

If we are in relationships where we feel more alienated than connected, we must work to communicate in a new way. That opens up our world to connection and bonding. We must not allow our fear to bring alienation into our communication. If we feel separated from others we wish to connect with, then we must find new ways to communicate for better connection.

Listening effectively is as important as verbalizing our own thoughts. To connect with others, we must feel heard, safe and understood. We must also extend this safety and listening to those we are hoping to connect with. Emotional connection and heart connection are special and rare. Physical attraction is easy. Souls connect and bodies attract. But once you have had a soul connection, a physical connection will never be enough. When we have this sort of heart bond, the connection cannot be broken. No distance can interrupt true connection.

We spend a lifetime understanding us and knowing how each part of the connection puzzle works. Every connection is significant. When one connection is severed, it affects all. We behave according to our connections, sometimes consciously and other times unconsciously. Everything is connected with one thing leading to another. Our light can go out, but another can bring it back. This is connection.

Our thoughts lead us to much pain. When we are able to neutralize our thoughts, we can take our power back. In our consciousness we can learn to see our thoughts, but not to assign goodness or badness to them. The reality is, our thoughts are not definite and not reality; the meaning we assign to them is the power they have over us. The meaning we assign to thoughts about the people in our lives influences us towards these others, sometimes in good ways and sometimes bad.

When we monitor our thoughts, we remove the conditional thinking that brings us pain. We are conditioned by life to think in particular ways, but if we are able to let go of our conditioning, we will find peace. After all, much of our early conditioning is applied by people who are biased in many ways, and they pass these biases on to us.

So much of our days now are spent on social media. It steals our joy and takes us away from practicing peace. It influences us toward accepting or reinforcing biases which are often not our own. We become anchored to others' conversations and thoughts, which often hold us hostage to our pain. We become captive to doctrines that bring suffering.

The way we work our thoughts to justify our pain is at times detrimental to our well-being. When the pain in our lives becomes intolerable, we tell ourselves stories to help us manage. Sometimes, however, the narratives are not accurate. Intentionally noticing our narratives can alter our perspective in ways that can bring us peace.

So much of social media is about planting thoughts in our mind. When we are in a vulnerable state, we don't question these thoughts. We don't pay attention to how they are affecting us. Are they adding to our lives or subtracting from them? We can learn to interrupt our thoughts and replace them with those that build rather than tear down our lives.

Taking the functions of our mind and using them in our favor is taking our power back. Our unconscious processes deliver pain and suffering, but they can also deliver joy. Being mindful of what we prefer and what we instill in our minds is essential. It's true that our brain works over 90% unconsciously, but we can bring it into our consciousness and reap great benefits.

Decode your thoughts and the meaning you attach to each thought. In this way, you can choose to let go of pain. There is an important distinction between consciousness and unconsciousness, and it determines the experience we are having in our lives at any given moment. We are often lost in thought, and many of these thoughts include worry and anxiety. These thoughts contribute to a diseased thought process which disconnects us from peace. Until we become intentional, we will be saturated with thoughts that take over our behaviors and mindset.

As humans we have the capacity for abstract thoughts. We can interrupt the connection between thoughts and suffering once we notice the thoughts. So often we think without even realizing what we are thinking. Becoming aware and not reacting, judging, or resisting is how we begin the process of positive change. Realize the anxiety you feel is not causing the pain so much as the reaction it causes. This is reframing our feelings into positives—which can make our lives better.

Here's an example. When I speak to groups, I might get nervous beforehand. The story I tell myself around this nervousness is what matters to my success in delivery. I can say this is a *good thing* that is giving me energy and motivating me to do well. It reframes the entire experience of anxiety into one that will assist my effective delivery of my message.

The way we frame our experiences and thoughts about them is significant. Let go of resisting certain feelings. It is a path to peace and diminishes pain. We react, and this complicates our experiences. It entangles us in a web of untrue narratives we connect with our feelings.

For example, if I am in line at the grocery, and someone cuts in front of me, my thoughts go to a dark place, and I can become angry. If I can wonder what is going on in this person's life and find gratitude that it is not going on in my own, then I can reframe this person's bad behavior. It won't affect me so negatively and impact the remainder of my day.

This is how we build compassion through our thinking processes. If we have a problem, we can look at it and determine what part of it we can fix, then we can move to do that and let the rest go. If we can't fix any of it, we can disconnect from the problem and go on about our day.

Life is full of problems. Remembering that we are not the only ones in life with problems and that everyone we bump into today has a problem can bring us peace and positive adjustment. We all wish for fewer problems at times, but we don't get to choose. We simply have to not allow our thoughts to make us suffer more than is necessary. The story we tell ourselves is based usually on past experiences. But is it bringing us pain to bring our past into our future, or upon analyzing it, is it best to let it go?

Eliminating the negative thoughts and choosing which ones we entertain is the goal of mindfulness. Becoming indifferent to our thoughts is a superpower for us all. We can choose more compassionate and empathetic thoughts through skilled, intentional behaviors. The script we reframe can keep us at baseline happiness.

CONNECT & ENGAGE: *Think about a problem that is weighing on your mind right now. Will this matter in six weeks, six months or six years? Use this timeline to reframe your thinking and come up with a rational plan to deal with the pain.*

CHAPTER EIGHT

Empathy

When people talk, listen completely. Most people never listen.

—Ernest Hemingway

NOT SO LONG ago, I sat in a session with another therapist who was working with a young girl who was sharing her feelings of never being heard or understood by her parents. The girl was clearly stressed, and said she simply felt like she was invisible in her own family. She felt there was no real feeling that they cared about what she thought or felt.

As she finished her story, the therapist's cell phone rang. She took it out and answered it. She apologized and said it was a delivery man, and that she needed to set up an appointment with him for the following day.

I watched this girl's face fall, and with it, her spirit. Once again, she felt ignored and unheard. It was truly one of the saddest things I have witnessed with a client. Clearly, she needed empathy from this person who was ignoring her feelings and needs, just like her family did. She was already in a state of overwhelm before this occurred. She left that appointment that day and never returned to therapy.

A central belief is that we move toward those who treat us as we are accustomed to being treated. It was interesting that she chose a therapist who treated her much like her family did. It was her comfort zone, even though it was a painful one.

Empathy is so important to relationships and the direction they take. Being available and helping the person opposite you by showing that you care and are there for them is crucial. In many ways it is a lost art. Responding with intentional empathy is critical. We must hear others and be considerate of their pain and the story they share. Doing so makes our world more user friendly. Showing up and caring is everything; this is empathy at its best.

Empathy is present in all of us, but often we have to awaken this part of our hearts. Perhaps we have been too distracted to acknowledge it, or we have buried it deep in order to feel safe. For many, vulnerability is not safe. For some, it is part of self-discovery. It is an important component of the journey of life. As we discover empathy within ourselves, we realize it was there all along, waiting to be discovered and bringing us into our wholeness. It is like being out in a snowstorm with cold hands only to discover our gloves were in our pockets all along. They were simply waiting to be discovered and placed on our hands to bring us warmth.

The process of practicing vulnerability is painful, but we can learn to develop courage and move closer to the pain. Typically, we want to run and distance ourselves, but as we move closer and become comfortable with the uncomfortable, we realize our own strength in this healing process. We have to become okay with going against the natural processes. We have to stay and face the truth—even when it is not pleasant for us.

As we grow on our own personal journey, and clear away the distractions, then we can move toward our need to share this with others. We can learn to be generous with our compassion and joy and give it to others. We can give that thing we most wanted for ourselves.

Empathy is found when we are not found manipulating it, but simply letting the experience be in us, both the good and the bad feelings. Once we are able to be empathetic, we become less reactive, and things become clearer. Even the hardest situations become workable. We can transform all of our experiences through empathy.

Empathy cannot be learned in a vacuum. It takes interaction between people and in-person connection. We cannot get this via smartphones or other electronic devices. We must first realize what is compromising our empathy. Researchers believe overconnection to electronics is one element that lends to low empathy between individuals.

Empathy is becoming a lost art in our society; so much so, that some large companies are bringing in experts to train their upper management in empathy. They realize the important role it plays in our world. Until we can be mindful of the other person's pain, we cannot truly join with them.

It is crucial to society that we exercise empathy, and understand others' emotions and their perspective. We learn this skill from in-person interaction and by picking up emotional cues from others. This is a learned skill.

We have mirror neurons in our brains which cause us to mimic the actions of others. But when we are interacting through electronics, we don't have this advantage. We can't pick up on others' moods or emotions. Think about when you see someone yawn and then you yawn. That is your mirror neurons at work. Joy, pain, and laughter can all be mirrored through personal interaction. These are emotions which are not relatable through devices.

To learn to be empathic, you must learn to observe others. Digital distractions hinder us from doing this. Taking the time to notice what others are doing is a way to build the empathy muscle. Put your phone down, listen, and make eye contact. Take time to genuinely care what the person opposite you is feeling. Be curious about their well-being. Learning more about others is a key element in building empathy.

When we are distracted by our devices, we lose the ability to walk in another's shoes and be connected through empathy. We miss out on receiving empathy and giving it. Loss of in-person connection is making empathy extinct. This lends towards narcissism and a world of self-absorption. The more connected we are to our devices, the less connected we are in person, and we become isolated and socially incompetent.

Face-to-face interactions can grow your empathy in ways that social media cannot. When we are able to hide behind a screen, we become keyboard warriors. We lose sensitivity to how our words affect others. People say things online that they would *never* say in person.

One way to build empathy is to put your device down and volunteer to serve others. There are many in-person opportunities to give of your time and energy. Connecting in this way is as beneficial to you as it is the person you are serving. Our world becomes a better place as we become more connected and engaged.

Good connection involves an internal connection through one's feelings. We are fully human when we can empathize. The chasm caused by no empathy is the greatest disconnect in human relations. The person who cannot offer empathy is not able to connect, and this is a travesty, as it serves to isolate the individual from the thing they most need—connection.

Fortunately, empathy is a skill that can be learned. We all have the capacity to be empathetic at different levels. We can practice this and master it. The more we practice, the better we become at being empathetic. Once you can learn to authentically feel for another, you become more genuine and forge stronger connections.

CONNECT & ENGAGE: *Think about how you react when you are around others. How might you become more tuned-in to what they are feeling? How might you pause and put your own needs aside to become more engaged and connected with the person in front of you?*

CHAPTER NINE

Stress

Happiness is a choice. You can choose to be happy.
There's going to be stress in life, but it's your
choice whether you let it affect you or not.

—**VALERIE BERTINELLI**

WHEN MY SON was in the first grade, he came home from school every day with a list of items he needed for the next day. Sometimes it was just one small thing, and other times it was something bigger. For example: flip through your family photo albums, and find five pictures yourself with five different aunts—each in a different colored shirt.

It seemed like this teacher had a case of the passive aggressives. I could make no sense of her daily "must have tomorrow" list. As a parent, afternoons were busy. When the kids came home from school, getting their homework done, taking each child to their scheduled activities, coming back home, and getting them fed and bathed was such a well-timed dance that adding this one more chore would often tip the scales of stress for me. My children's father was rarely home to help, so this of course compounded my stress. Single parenting is not for sissies; I learned that quickly.

It got to a point that I would begin anticipating this "list" before my son even got off the bus. I spent my afternoons planning how I was going to manage the scavenger hunt I was forced to participate in every afternoon by this seemingly clueless teacher. He was my first child; I wanted to get everything right and felt one parental slip would surely be catastrophic for his future. I can remember thinking, *Please, for once, can he just get off the bus and us not have this stressful to-do list?* I just wanted to be together with my kids and enjoy those few hours before bedtime.

As a mom, I was in a constant state of overwhelm as a direct result of what seemed like thoughtless, last-minute assignments. If I was under this much stress due to my son's assignments, I wondered how they were affecting him? It seems like stress starts earlier for our kids now; our choice of schools and activities for our young children can be a direct cause of stress.

I can remember wrestling with the decision to send my kids to an elementary school which required long assignments for nightly homework. It left no time for them to climb trees and simply be children. It felt like I was introducing stress into their lives—at a time when they should be learning anything but stress—through my overscheduling and commitments.

Shouldn't childhood be a little more carefree? Isn't that what it's all about? Won't there be time for all of these demands on their time when they are adults? It felt like every day they had to give up an important part of being a kid to stay in this school. I know the first eight years of life are crucial and when kids' brains are developing. How would they be affected by so much stress during this important developmental stage?

Never before have our adolescents been under such stress. The suicide rate for teens is at an all-time high. In fact, it's one of the leading causes of death for this age range.

The age of screens has added pressure to their lives in ways we don't recognize.

Are smartphones wrecking this generation?

Some researchers believe this generation is on the brink of mental breakdown due to their overindulgence in smartphones. A few decades ago, teenagers were spending time at the mall, hanging out together at the movies or at the pool for the summer. Now we have so many teens reporting spending time alone in their rooms with their smartphones. Their summer companions are their smartphones. Their world is a world someone else is living.

Some teenagers report liking their smartphones more than they like actual humans. This is a huge problem. As humans, we are wired for connection, and when we don't get it, stress becomes a major factor in our lives.

Just last evening at a restaurant, I watched a young family sit down for dinner and give each of their young children, under the age of ten, an electronic device to play on during dinner. There was almost no interaction between the kids and adults throughout the meal. Mealtimes are an optimal time to teach kids how to interact properly with other humans. If they are learning this early in life, it will not be a lost art when they become adults.

It's another stress builder for them as they grow older and try to socialize and not be awkward doing it. These are all skills we learn early in life. When parents are plopping kids in front of screens rather than helping them learn to connect and engage, they suffer later in life with isolation and loneliness. Not surprisingly, isolation and loneliness significantly increase premature mortality rates.

We are noticing shifts in the emotional and behavioral states of people, and this can be connected to how we are spending our time. The experiences we have or don't have every day are significantly different than a few decades ago when we had less technology. People in every corner of the world are majorly affected by smartphone use.

Stress levels are higher than ever, and many feel that the rise of smartphone use and social media are contributing to this. More teens are not leaving home to spend time with friends, because they think they can do that on their phones from their home. But it is not the same. Much of the time spent in their rooms is stressful for them. They are constantly comparing themselves to others on social media. This increases feelings of loneliness and isolation. Consequently, in spite of their so-called connection, they are feeling more left out than ever.

Surveys show that people who spend more time on social media and smartphones, are less happy than those who don't. Less happiness can be linked to screen time, and more screen time can be linked to depression and anxiety. Putting down the screens and engaging in any screenless activity is known to raise happiness levels and decrease stress.

Technology is developing at a rapid speed in our world. In many ways it adds ease to our lives; however, we must notice also the ways it detracts from our lives. We must become intentional about monitoring our personal use of technology, as well as our loved ones' usage habits.

One way to do this is by being honest about all of the ways it has changed our lives—both negatively and positively. Awareness is everything, and once we are aware of the ways we can better connect and engage with others by putting down our technology, then our lives will benefit.

Also, the lives of our loved ones will be better with our increased and improved attention and presence. There is much to be said for teaching our kids early in life the importance of in-person relationships and being present with the people in our presence, rather than distracted and disengaged.

I always know where my digital device chargers are, and I rarely leave home without them. If I do, I freak out at the thought of being isolated from my online world. Isn't it interesting how we keep our phones charged, and yet our bodies are often running on low battery mode and we ignore the signals they send us?

Stress and anxiety are our red flags, but we pay no mind. We don't take time to unplug and restore and recharge our batteries. If you are feeling stressed and depleted, try reading a book, or having coffee with a friend. It is so much greater than scrolling and liking. Physiologically, it can be the recharge we need to restore our bodies and ready us for the next big task.

We must notice when we are taking better care of our digital devices than we are of ourselves—particularly our minds and our bodies. Digital devices can make our lives easier, but when we cross from average usage into obsession, we become inundated with information and experience stressful information overload. This releases toxic brain chemistry and increases depression, anxiety and stress.

The thing you move toward when stressed is important to note. How does it make you feel? Do you feel more anxious and stressed after engaging? If so, it's time to change your habits and make better, healthier choices.

CONNECT & ENGAGE: *Think about your level of stress right now. How would you rank it on a scale of 1 to 10? Now that you know your stress levels, what can you remove from your life and what should you add to your life to make it better?*

CHAPTER TEN

Intuition

Intuition is seeing with the soul.

—DEAN KOONTZ

WHEN I WALKED into a nail salon, the receptionist was on her cellphone. After I spent a few minutes standing there waiting, she hung up and began talking to me. Almost immediately, her phone rang again, and she answered it. She left me standing there waiting while she shared her plans for the day with a friend. She seemed to realize that a customer was standing there so she briefly shifted her attention back to me. She told me to pick a color and then quickly got back on her phone.

I was starting to have a bad feeling about this manicure.

When it was finally my turn to have my nails done, the man was amazing—not for how he worked on my nails, but for how little attention he gave them while seemingly being engrossed in a conversation on his phone. His focus was definitely elsewhere.

He asked me whether I wanted round or square for filing, while his phone was pressed between his shoulder and his ear. I told him I wanted round, and he filed them square. He asked me if the water was too hot and when I said yes, he made it hotter. Shortly it became very clear he was quite distracted.

I finished my horrible manicure and got out of there as quickly as possible. I made a vow never to go back.

We have to realize that our time is valuable, and we honor ourselves by expecting undivided attention. It's not unrealistic to walk into a place of business and expect the salesperson's undivided attention. They are there to provide a service, and if they are unwilling to do so, it should send up a red flag. Don't be the person who can't ask for what you need—undivided attention—and don't be the person who doesn't offer it.

Block, unfollow, and delete. These should be our go-to buttons on our digital devices. We have all of the power. When someone makes us feel unimportant or negative, we can block them. Why expose ourselves daily to their negativity? I have found that many on social media use their platform to flood the world with their negativity. If you are consuming it, it's garbage in, garbage out. If I begin my day reading their gross displays of misery, then my day tends to follow in their direction. It's a perfect reason to use the block button.

Use your intuition to only let in the good. You know whether someone's posts are going to make you feel good or bad. That's one thing increased time online has taught us. So it is your choice whether you grow weeds or flowers in the garden of your mind. The care you take of it is your responsibility. You would not put your flowers in the hot sun with no water, so don't put yourself in the way of another's negativity online. Be willing to show up for yourself in the best way possible by hanging up and disconnecting. Stop engaging with the meanies on digital devices!

How many times have we looked back and said: *if only I had listened to what my intuition was telling me*. We each have this still, small voice that guides us. Yet somehow at times we become disconnected from it, and we don't listen. This can lead to painful consequences. In our world today, there are so many distractions that keep us from paying attention to or hearing our quiet voice that wants to lead us down the right path.

Dr Bessel van Der Kolk, an expert on trauma, says we must be "free to know what we know, and feel what we feel." How can we do this in the age of distraction? How can we notice our intuition when we are so busy looking for the answers on social media and video games? These distract us from ourselves. We can't know what we know if we are too distracted to listen to our intuition.

Most of us have no idea of the internal struggle going on inside us because we are so plugged in externally that we just don't notice. The common pressure to keep up with others on social media keeps us looking outward. The constant interruptions and expectations to be available always via texting and email keep us looking downward. How can we ever expect to have time for ourselves when we give everyone else 24/7 access to our time?

It seems like a generational challenge always exists, and for the current generation I would say it is finding "me time." Me time is time to be disconnected and unplugged from technology and plugged in with self. It is quiet time that is critical to a healthy well-being.

When we are constantly juggling the effects of the dismal news, which is always in the background, and the never-ending internet distractions, we can't realize the importance of our own thoughts and voice. They get lost in the barrage of overstimulation due to tech.

The self-absorption we see from frequent selfies and non-transparent self-marketing on social media surely disconnects us from ourselves. What we post isn't truly us; it's the us we want the world to think we are. We begin to believe the stories we view on social media. They convince us that everyone's lives are over-the-top great. Everyone's except yours, that is.

We become dissatisfied with self and stop listening to our own voice. We lose touch with us. It's difficult to not isolate when we feel less than and out of touch with our true selves.

Don't quiet that inner voice. If it's been a while since you've heard it speaking, it may be time to put the device down and get back in touch with the real you. Take a walk outside. Sit and look at nature. Journal about what you are thinking and feeling.

Learn to be comfortable in silence. It's the only way your intuition will be able to speak to you. But when it does, it will become your steady guide.

CONNECT & ENGAGE: *Think about your intuition. How connected are you to your inner voice? Is it quiet and barely audible? If so, where might you go and what might you do to help it speak up? Make it a point to carve out a couple hours to just go be by yourself and listen to what your intuition has to say.*

CHAPTER ELEVEN

Community

*The most terrible poverty is loneliness and
the feeling of being unloved.*

—**MOTHER TERESA**

LONELINESS IS A top indicator of early death. It
contributes to health issues like diabetes and heart disease.
As Maslow points out in his hierarchy of needs, being
connected and engaged with others is a basic human need.
It is how we are wired.

Isolation is not our friend. It affects us in many negative
ways. Yes, sometimes it is nice to have "alone time", but it
should make up a small percentage of one's day.

Being connected and engaged makes us happier and
more at peace in our lives. For most, having family and
friends is the most important part of their lives. Feelings
of loneliness affect all ages and are not gender specific. The

elderly and adolescents are most likely to be affected by disconnection. We know the limbic system and the "fight or flight" response are affected by loneliness, just as they are when a person faces a dangerous situation.

Physical health is affected by loneliness, but mental health is also a factor in isolation. If you are lonely, you are more likely to develop dementia.

Loneliness is also connected to arthritis. Our immune system is compromised when we are lonely, and we make more stress hormones. Elevated stress hormones negatively affect our sleep. We know that loneliness is a chronic stressor, which can affect one's physiological well-being and mental health.

Loneliness is thought by some to be a public health crisis. It is a universal emotion. We know that it can be addressed through connection and engagement for those in isolation. Increasing social interaction is as significant to mental health as good dental hygiene and exercise are to physical health.

Decreasing loneliness can improve heart issues, depression, stress and anxiety levels, memory problems, and addiction issues. Recent research supports addiction being connected to disconnection. Addicts in recovery stay in recovery longer when they remain in connection with others. A major factor in addiction relapse is feeling disconnected.

Many who suffer from loneliness describe it as feeling unwanted and empty. This way of thinking causes them to isolate and makes connecting and engaging with others difficult. For some, it is a constant state of mind. Many report feeling alone in a room full of their peers. It can be connected to our genetics as well as situational factors. Those who struggle with self-esteem issues often report feelings of loneliness. They feel they are unworthy of being connected with others. This story they tell themselves often perpetuates the problems.

A helpful way to combat loneliness is to become connected and engaged. First, we must alter the maladaptive stories we tell ourselves and change our thoughts to be more self-supporting and positive. Next, we must limit the time we spend in front of screens, and instead spend time making in-person connections. Having three close friends is sufficient to eliminate loneliness and improve our health and our state of mind. It is very hard to maintain more than three very close friendships. We get spread too thin when we try to be truly intimate with more than three.

Research has discovered that loneliness can be contagious. If we have a lonely friend, we are fifty-two percent more likely to become lonely ourselves. We can make conscious efforts to make changes that help us overcome loneliness. Only when we become aware of loneliness can we make changes to alter our lives to be more connected and engaged. Being intentional is half the battle.

When we are aware of the effects that loneliness has on us and our physical and mental health, we can make improvements. We can join community service groups and practice hobbies we enjoy. These give us great opportunities to meet others with similar interests. We learn to foster new friendships and build more social interaction into our lives.

Much research has been done which proposes that *community* in America has dropped off significantly. The result is less trust in society and more individual isolation. People are less engaged in politics, and most people don't know their neighbors. These are all factors in the growing issue of loneliness. Again, we are experiencing fewer close relationships and more followers on social media. A recipe for isolation, this unhealthy and unrealistic comparison leads to disaster. Kids are connecting more online through gaming but less in groups and in-person.

Just this week, I slowed down from work and travel and spent time around a community dinner table with friends. I felt my stress level immediately fall, although my circumstances had not changed. Simply being in community with loved ones contributed to my higher spirits. It's amazing how liberated I felt from my stressors after just one evening out with friends.

We can shift our thinking to have higher expectations. We can expect better outcomes when we engage with others. Where you stare is where you steer, so be mindful of the ideas you tell your mind. The mind believes everything we tell it. Then it moves in that direction, whether good or bad. So we must try to tell it only factual and positive information. Focusing on the positive can shift our attitudes and our relationships and benefit our lives greatly.

We can make friends with our loneliness and use it as a motivator to get out and make friends. Courage of this sort can improve connection and encourage us to thrive in new ways. Our loneliness can make us aware of changes which result in our being happier and stronger. In this way, loneliness is not harmful; we can use it to face new challenges and bring life improvements. Viewing loneliness as an indicator of how you are feeling will bring about higher emotional well-being and productivity.

The way you think about your loneliness will be a deciding factor of how you manage it. Loneliness will be more harmful if it is not something we choose and is out of our control. If we are able to change these circumstances, we can move away from isolation and towards connection. People report feeling their lives are more meaningful when they are better connected. Rather than deciding something is wrong with you when you are lonely, look at it as a barometer for how engaged and connected you are. View it as an opportunity to grow and learn.

Loneliness imprints on our brains in ways that can help us handle it differently in the future. In a sense, it inoculates us as we learn and grow from the experience of loneliness. Being lonely and overcoming it can actually help us avoid it in the future.

It can help you feel less anxious and more hopeful for your future when you are connected with others. We are all connected in some way. Reaching out to others and building strong friendships based on these similar connections can be life saving for some, but life improving for all. When we have experienced loneliness personally, it can help us see the loneliness and disconnect in others and move toward them in hopes of connecting.

I am excited by technology, but I believe we are allowing it to take us to scary places. These can be lonely places, far away from community. Our digital devices are changing *who* we are, but also *how* we are. We are becoming desensitized to being disconnected. Making eye contact is a lost art. We are learning to be together, but not present.

I have observed introverts at cocktail parties removing themselves from the party mentally and emotionally, while still being there physically, by going into their phones. We are getting used to being alone while together, and this is becoming acceptable.

We are constantly connected through digital devices, but we are disconnected in life. We are becoming okay with observing each other at a distance while sharing the same room.

We are lonely because we don't have conversations. Conversations online allow us to edit and delete our conversations so we can control them, but in-person communication is not so clean. In person, conversations require undivided attention and presence. Sadly, we just aren't able to be as present anymore. Being in a conversation we can't control is scary. Conversations in person are messy and involved, but that's how you build community.

Texts and tweets do not add up to real conversation or real connection. They are great for sending short messages, but we can't know or be known through texting. The consequence is a false connection and a feeling of loneliness. Most people would rather text than talk. This is the new trend, and it is causing loneliness. We don't feel heard, so we move toward digital devices to comfort us.

We are a vulnerable world. We want connection and escape from loneliness. Our expectations for technology are greater than those for our loved ones. We have taught ourselves to be afraid of intimacy. We have desensitized ourselves from the expectation that we can be connected and heard and do not have to be alone.

Have you noticed when you are in a group who doesn't know each other that they can't be alone together? They all take out their phones and begin to text. They are right there with a group they could engage with, but instead, they reach out to others through their digital devices to feel less anxious. Ironically, this way of always being "connected" is really just causing us to be lonely. We act as if solitude is a bad thing. We avoid it.

We can reconsider and become aware of how we are using the devices as the things that take us away from real world connections. We are substituting digital devices for evenings with our friends. We need to look for ways to use our digital devices to make our life better and more connected.

We self-sabotage when we isolate and carry negative stories in our minds about ourselves. This negative narrative keeps us from connecting with others. By not trying to eliminate loneliness from our lives, we engage in self-sabotage.

When you engage with others, you know you have found a good relational fit when your energy is lifted in their presence. We can learn to honor ourselves and our feelings by making healthy choices and discovering what makes us happy in our community with others. Being in touch with how we feel in community with others is a mindful way to connect. Being conscious about the way we feel in our relationships with others can contribute greatly to our happiness.

Community offers support in the form of listening. Sometimes it is in saying nothing, but simply being present with us while we figure it all out. We can stay in healthy community with each other by offering to help with tasks, checking in, and simply keeping company with them regularly. Others will learn to depend on you and count on you always because they know you show up regularly. This is community.

It's hard to be in community when you are constantly making excuses because they are barriers to connection. So let go of the past and get connected to others. This is how you change your future and combat loneliness. We are all imperfect people, but we can find connection and healing through community.

CONNECT & ENGAGE: *Think about your level of community. Where is it strong? Where is it weak? How can you better engage with others to form stronger and more beneficial community?*

CHAPTER TWELVE

Connect with No

No snowflake in an avalanche ever feels responsible.

—Voltaire

IMAGINE THE FREE time you would have if only you could learn to say no.

No, I won't make the cake.

No, I can't attend the seminar.

No, I won't answer that call during dinner.

No, I won't answer my emails except for once a day.

How would this free up some time for you to do something you enjoy or connect with a friend in person?

When I sit down to write this book every day, I can be distracted by the pops-ups, the emails, and *Oh, I need to check in on Facebook.* There are so many choices other than the one I need to focus on, and they are *all* thieves of my time.

I have learned to say no to anything other than the goal in front of me. Is this easy? Absolutely not. I can rationalize with myself why it would be nicer to catch up with friends on Instagram than to finish writing my page quota for the day. It's not hard to find reasons not to be productive. And all of the reasons look good.

But prioritizing and remembering why we set the goal in the first place is essential. Some days we don't feel like it. But if we wait for the motivation, we will grow old waiting. You have to learn to say no to everything that is not part of your goal.

Do you feel like you have to immediately answer every email that you are sent? What about every instant message? Every text? Why? Will the world end if you don't reply? Think about it this way, if you allow all of these constant cyber world distractions to invade your life, you will never reach your goals. You could be checking and returning emails 24/7 and never living your life. Every correspondence does not deserve a reply. Just like every request of your time does not deserve a yes.

We must pick and choose wisely.

Sometimes we have to learn to tell ourselves no. It's not just others that can distract us from our purpose. How many times have you been running into a meeting when your phone rings? When you stop and answer it, your five-minute talk turns into a twenty-minute phone call. You walk in late after keeping others waiting who actually scheduled time with you.

We don't always realize the obstacles our digital devices cause to our daily schedule to keep it flowing smoothly. We don't properly prioritize texts and phone calls that should wait. We don't need to be accessible to all twenty-four hours a day, seven days a week. They will live. Our anxiety tricks us into thinking that we're urgently needed, and then our actions double our anxiety.

We could all add more time to our day for important things, things that matter, simply by being aware of our digital device usage. We do have a choice. We can choose not to answer the text right away. We can answer our email at set times. We can create boundaries; they truly fix almost every problem. Boundaries help us resist the urge to answer our digital devices when we should be engaged and connected elsewhere.

You get to be in charge of your digital devices. Don't let anyone else's expectations pressure you in this area of your life. It throws your limbic system into high alert to constantly be on call and maxes out your anxiety levels. Bring calm into your life by monitoring your device usage, and learn to say no. Take back your power by taking control of your device usage. Have a designated time to be connected digitally. At all other times, live presently in real life.

I decided to do a self-experiment, so I left my device at home when I went to lunch with a friend. Honestly, it was really hard. I felt so strange. I reached for it a couple of times and felt a shudder when it wasn't there (phantom vibrations). Driving off without it truly brought up some anxiety for me.

I realized through this reaction that I, more than anyone, needed to separate from my phone. This separation anxiety I was feeling was very real and a huge red flag for me. I began to challenge myself to leave my phone more and more. Now it feels natural to go to lunch and give my friend my full attention, and I feel horrified that I would consider ever not doing so.

What was I thinking before?

My son busted me one day. He walked through our house while I was working on my laptop. He jokingly told his girlfriend Abby, "Isn't it funny that my mom has this platform about digital device usage, and yet she is always on hers?"

It's true. Kids say the darndest things. Do you listen to your kids? Do you hear what they are saying, and does it ever cause you to pause and wonder if just maybe they are onto something?

If you feel disconnected, put your phone away and stop comparing yourself to the folks on social media. Go out and have some real life in-person experiences. Be fully present for them. Say no to anything less.

Saying no often means overcoming your fears. We all have them. Fear is the most common human emotion and drives much of our behavior. But it doesn't have to be that way. We can overcome our fear and say no.

When we are able to say no to fear, we can become better connected to ourselves. When we develop good boundaries, we can overcome rejection and not be fearful of speaking up for what we need. We can be afraid and do it anyway. This requires being vulnerable and connecting with our scaries, and sharing ourselves with others—especially when we are afraid to. Our relationship with fear changes when we find the courage to say no to fear.

We can let our past painful stories go and make a new and better ending. We can learn from the past and say yes to our brighter and more peaceful future. Once you understand that you deserve better, you will become better at saying no to all that does not provide you with what you need for you.

Remember when you say no, you grow.

CONNECT & ENGAGE: *Think about how you establish boundaries. Is it easy for you to say no? Why or why not? What are some small areas of your life where you can take back lost ground and begin saying no?*

CHAPTER THIRTEEN

Go Through, Not Around

Endurance is one of the most difficult disciplines.

—UNKNOWN

AS A THERAPIST, one of the concepts I help my clients work on is *not* avoiding their problems, but instead, going through them and all that entails. No, it is not fun, but in the end, it is the only healthy way to live.

Today we have so many distractions which can hinder us from going through our pain and avoiding problems that we become numb to their very existence. We know that we can surely grow if we go *through* rather than *around*. Much of our pain occurs in relationship; therefore, we must heal in relationship. We have healed when we can truly say we have a real connection with another person. This is often the shift we need to take us from isolation into community.

So why do we avoid this? Why do we go to great lengths to go around rather than through our problems? Well, most of us don't enjoy pain, it's true, but if we can realize on the other side of the pain is peace, then we will be more inclined to take the through route.

If we are going to come out the other side new and improved, then this might be an exciting perspective. Scary, yes. Change is scary, but so is staying stagnant. The question many people never stop to answer is, *Why do I keep going through the same negative experiences?* The answer is because you are not willing to go through the pain of being uncomfortable for a small window of time in order to find peace for a lifetime.

If nothing changes, nothing changes. And if nothing changes, you stay stuck.

Being pushed out of our comfort zone to step into truth is a difficult but worthwhile thing. If you listen to the stories of the most successful people, they all had to give up something or leave something behind in order to reach their goals. Much of the time, they were not conscious of the thing they needed to let go of or move past. When the opportunity came to leave this problem in the dust, they didn't realize this was what was happening. Only in hindsight did they have this important revelation. It was pivotal in their lives and their success.

So what are you allowing daily distraction to keep you from going *through*?

What ways might you grow if you engaged in your life and connected with the things that would grow you?

How are you staying in a funk when you could be focused on something more positive and going about achieving it? Here's a tough question: *Do you spend hours every day scrolling, swiping, and liking others' experiences, rather than inventing and living your own?*

Today, can you accept the challenge to limit your screen time and go through something you have been going around? Can you limit the distractions and engage with others in new ways that will grow you?

Obstacles are often nothing more than opportunities.

When I work with couples, I help them to shift their focus from the feeling that obstacles are bad and must be avoided, to the feeling that obstacles are great ways to build your personal strength and your couple confidence. Going through the tough things only makes us stronger. There are not many, if any, exceptions to this rule. When you shift your perspective to embrace obstacles and go through them, you will notice positive changes in your life.

Those storms that you thought came to destroy you, perhaps they were merely clearing a path *for* you. Imagine that. I don't know many who don't look back at their life, see the hard things, and say, "I really grew from that."

That job I lost, it caused me to go back to school and gain more training for reemployment. Now I love what I do, and it doesn't feel like work. Or the ex-wife who finally realizes love after her abusive husband left her. Now years later, she is married to the man of her dreams and living her best life. She knows she would have never left the bad marriage without the abuse, but now she is grateful she left

When we reset our goals and have confidence, we have won before we have even begun. Confidence is everything. The mindset we take into our troubles with us determines how we will come out of them. It gives us the courage to navigate even the hardest of paths. We know that if we avoid the hard things to not feel uncomfortable, then we will always be at war with ourselves. We will not be living our best lives, and we will slowly implode internally. We have to be afraid but do it anyway. Courage is all about noticing the fear and doing it anyway. The best thing you will ever do is to believe in yourself.

The next best thing is to surround yourself with people who will lift you up and encourage you as you walk through your obstacles rather than around them. Look around you and decide if the people around you are adding or taking away from your life and plans.

Sometimes these sorts of questions lead to determinations that are difficult to carry through but necessary for growth. You may have to burn some bridges that you don't need to cross again. This will make it impossible to do anything but move forward. How scary is that?

This is why you need strong people to encourage you. You need givers, not takers. And you need to allow these negative things that have happened in your life to inspire you to walk on and move into a better place. If you made a mistake along the way (and we all do), forget it. Remember the lesson instead. There is no need to dwell on the mistake. Don't look back—you are not going that way. Look for stars in the darkness and rainbows in the rain.

Your best teachers are your past mistakes.

Don't go around them, go through them. The view from the other side is incredible.

CONNECT & ENGAGE: *Think about a problem you've been wrestling with for a while. As you call it to mind, be honest with yourself about why you are avoiding it. Then make a plan to push through the problem. Envision what your life will look like once it is solved, and then move forward with courage.*

CHAPTER FOURTEEN

Avoidance

*Choosing to avoid uncomfortable feelings offers
immediate short-term relief, but avoidance
can lead to long-term consequences.*

—AMY MORIN

ACCORDING TO ARISTOTLE, we are what we repeatedly do. We are creatures of habit, and those habits define us.

As humans we are hardwired for behaviors that form habits. In fact, after doing something for just thirty days, it becomes so automatic that we don't even notice that we are doing it. It just happens. We don't think too much about it, we simply unconsciously move towards it. We don't consider its effect on us.

According to research, a large part of our behavior becomes unconscious. Have you ever gotten home and wondered if you stopped at the stop sign at the corner? Of course you did; you have just driven that way so many times that the drive is an unconscious habit.

Habits become habits for many different reasons. Some habits are formed to avoid painful thoughts or uncomfortable feelings. If I come home at the end of the day and swing through Burger King to eat a Whopper Jr., it might take my mind off the stress of the day for a short time. Once I realize this, it may become an unconscious and soothing habit.

Sometimes we do need to look at our habits and wonder if they are helpful or hurtful. We can become so uncomfortable that we incorporate many different vices into our day to ease our pain. Internet porn, food buffets, sleep, or shopping can all become unconscious habits that we turn to in times of stress. All of these habits help us avoid the truth, but they become a thing we need to use for escape. Addiction is an illness of escapism, and avoidance is a much-used path of escape.

Imagine all of the things we avoid when we have our heads down in our devices. Also, imagine what we miss. I am always amazed watching people scurry to take pictures of once-in-a-lifetime experiences so they can post them to social media. In the process, they miss the actual experience itself.

Sometimes though, we sit on the sofa liking and swiping as the day rolls by unexperienced because we consciously or unconsciously don't wish to engage in our lives. It is easier to live in our heads or through others' lives than to put down our devices and actually engage in the day in front of us.

We don't realize that living vicariously through others is not authentically living. It is merely existing.

We begin living our lives by first facing them. Not all of the things we acknowledge can be changed, but certainly nothing can be changed that we don't first acknowledge. Timing is everything, so the "when" of facing the unpleasantries in our lives is significant.

Jung says, "People will do anything, no matter how absurd, to avoid facing their own souls." Immersing ourselves in our digital world to forget is another way of avoidance. This is temporary happiness, but it shortcuts the reality we are trying to avoid. We often make the choice to indulge in digital devices to avoid our depression, aggression, fear, and insecurities. We allow external indulgences to take us farther from our truth and our authenticity.

There is nothing like putting down your digital toys and all other external distractions and diving headlong into you. Be present for yourself in this mindful way, and leave the external world behind for a large portion of your day, every day.

Begin with minutes and build to hours. Then maybe you'll move to days without digital interruptions or distractions. It is such a good feeling when you detox from tech and live free.

It is so stressful just to have a phone close by because you anticipate the constant notifications. We exaggerate our importance in our minds and create a false reality that the world can't turn without us. Have you ever noticed that if you don't answer a call immediately and the caller has a problem, by the time you return the call they have likely worked it out themselves? Amazing! We often reach for our phones in avoidance of doing the work to figure things out; it is the easy way out to call someone and get them to do it for us. All the while, we are quite capable and can work through things solo—if we will give ourselves the chance.

When you notice you are over-occupied with your digital devices, ask yourself, *What am I avoiding in this moment?* Is there something I could be doing to be mindful and present in my life? How might I have an encounter or experience that will actually build self-growth?

Thanks to the internet and social media, it is easier than ever to avoid our problems and distract ourselves for hours online. Anxiety is at an all-time high, and the way many deal with it is simply to *not* deal with it. Avoidance.

We practice avoidance through binge watching and gaming marathons. Or we use our smartphone app to order up some tasty treats to be delivered right to our door. For a moment, it takes our mind off our anxiety. Avoidance.

There are many things in our lives we can't control, and therefore we are forced to go through them. Our response is directly related to the manner in which we get through things we go through, rather than around. We can't control the story others tell themselves about us or how they respond to us. We can't make others understand situations as we do. We cannot make others engage in behaviors we think are best for them, nor can we control how others feel about things.

When we begin to realize that there is much we cannot control, it forces us to go through reality rather than around it. We begin to let go of the need to define what happens or when it happens.

We avoid so much of the good stuff in life unintentionally by connecting with worry about what others will think of us. We have to remember that we all have a unique purpose, and we must never dim our lights to fit into another's world. When we dim our light, we become ill and depressed because we're being run by what others think of us.

Remember, people who are content with mediocre will not be excited about your quest for excellence. When I was living a ho-hum life and not doing much with it, it was quiet, and there was not much rumble from the stands. But when I decided to return to school and pursue my Doctorate, many ears perked up, and what they had to say was not all positive. I was astonished that not everyone wanted to see me better myself. Clearly, some felt threatened. But it was my choice not to make myself small in order to accommodate others. As Brené Brown says, "If you are not in the arena also getting your ass kicked, I'm not interested in your feedback." Don't dim your light for others; give them shades.

Our wish to climb the mountain reminds others of their shortcomings. They can only react with jealousy. For some reason, others make our wish to grow about them. It doesn't matter; do it anyway. We can be compassionate for those others while still reaching for our dreams.

The moment you move towards your purpose you will need to make some changes in your life. Some relationships may have to fall away. Some habits will need to be addressed and left behind. Addressing these things is critical if you want to meet your goals.

I have a friend who hired a Sherpa to guide them to the top of Mt. Everest. The very first thing the Sherpa said was that much of the baggage my friend had brought for the climb would have to be left behind. There would be no need for most of it, and no energy should be expended on the extra and unneeded baggage.

What is the baggage you will need to leave behind to get to the goals you have set for yourself? What is weighing you down? Stop avoiding, start addressing, and get to work.

CONNECT & ENGAGE: *Think about something that you've been avoiding. It could be a difficult relationship or a bad habit. Get honest with yourself, and make a plan to deal with it. Stop waiting! Get started today.*

CHAPTER FIFTEEN

Truth

*People don't want to hear the truth because
they don't want their illusions destroyed.*

—FRIEDRICH NIETZSCHE

HOW MUCH OF what we see on the internet is true?
More importantly, how much do people *believe* is true and
then act from that belief? What about our kids who are so
impressionable and believe all that they read; how does this
effect their decision making?

This one really sits heavy with me.

I read things on the internet and am simply shocked at
the absurdity and think, *Well surely no one really believes that
fake news.* However, they do. Anyone can write anything on
the internet. It's up to the reader to sort the truth from the
lies. When an unsuspecting person comes along and believes
it, they repeat it, and on and on the fake news goes.

It happened to me a couple of weeks ago. I saw online that Clint Eastwood had died, and I repeated it at lunch to a group of friends. Later in the evening, with another group, I repeated it again, surprised no one knew this. My friends asked me where I saw this, and I told them I had seen it online. They showed me that it was "fake news." Clint was still alive and kicking.

How embarrassing to be caught up in the scam and repeating it.

The Justice Department made indictments recently against eight people for a $36 million internet scam. It was the largest digital fraud operation ever to date. This is an enlightening commentary on how easy it is to scam people online. Bots seem to outnumber humans online now.

The way we gather metrics online is a perfect example. We can't really measure engagement accurately, because there are now "click farms" where hundreds of phones are lined up clicking on specific videos interacting as traffic. It drives views up but isn't connected to a person.

Fake people can post fake content and attract the attention of brand reps to get real money for their ability to be an *influencer*. We have a difficult time distinguishing the real from the virtual anymore. How much of what you see is photoshopped, and how many young minds believe it as truth?

It leaves one wondering how often we are lied to and scammed on the internet. What is reality and what is not? These days, it's hard to tell.

Every day in my office, I have the opportunity to help at least one person realize that just because someone posts it, that doesn't make it real. Rarely do people put their complete truth on social media.

Think about the things you share. We always try to dress it up to look a bit better than it really is. Some are more skilled at this fiction-making than others. When an unsuspecting person comes along and believes everything they see, their mood plummets dangerously.

We must realize that on the internet all is not truth, and all is not trustworthy. The internet is a tool many of us use to gain our news and to communicate with others. Research has shown that people over 65 are far more likely to repeat fake news than other age groups. There are several reasons for this, one being a lack of internet savvy. Fake news is now a big commodity, and the public can hardly keep up and not be fooled. Online disinformation is rampant. False reporting attracts eyeballs, and this is how many advertisers make money.

Fake news is not the only fake thing online today. There are also fake people. Anyone can be anything on social media, and much of the population will believe them. There are vile people who paint themselves pretty for social media and seem so caring and compassionate. They seem successful and smart, because well—you can be anything you want on the internet.

At some point we have to be realistic about what we are seeing and begin asking some critical questions. That nudge we get about wondering who this person really is, well, we should listen to it. Our intuition is a gift, yet so many times we tune it out and then find ourselves in turmoil later.

If an online influencer seems too good to be true, they usually are. It doesn't matter if people sit in the front row at church, or they are the first one at your door with a casserole when you are in need—if you have a feeling telling you otherwise, you'd be wise to listen to it.

We get so distracted by our digital world that we are losing touch with our internal intuition. We buy into the stories we see online, and many of them are just that, stories.

Those folks who pose around the Christmas tree all smiles, or from their ski chalet in Switzerland seemingly with not a care in the world, trust me, they have cares. Staged photos are not a true representation of real life. Comparing yourself to this is a disservice and holds you back from your own healthy living.

At the end of the day, the biggest obstacle you will have is you, based on the stories you tell yourself. You must trust your gut and not hesitate based on things you have seen online. These things can work against your dreams. Teach yourself to follow your instincts and put aside the chatter. Hear the small voice in you and follow it. Trust *you*. Don't be derailed by what others are doing that may not even be real.

I once had a lady confide in me that she got up every morning, reached for her phone, and began scrolling. Each day, she saw lots of what appeared to be perfect people and their perfect lives on social media. After thirty minutes of this she decided to put her phone down and go back to sleep. She felt *less* and told herself she was not enough. She believed everything that she saw online.

Part of our work together was my helping her to leave her phone in a drawer for the first two hours after she woke up. She learned to go about her morning, have quiet time with a cup of coffee, and then listen to something uplifting while she got dressed for her day. After that, she walked her dog. This new process set the tone for her day. She told me that it was much better once she learned not to begin her day scrolling. She used to feel overwhelmed before her feet even hit the floor. Now she's excited to start her day.

You get to choose how you begin your day and what you will do with your days. Create the life you wish for, and stop competing with the life you see others living online. We can learn to visualize the thing we want for ourselves and the truth we want to live. When we are in a state of cognitive dissonance, our thoughts are not in line with our life. This causes anxiety and other maladies.

As a therapist, one of the first things I do as people come to see me is determine where they are not living *their* truth. This is not conscious. We have to move toward our visualization, while at the same time being joyful in the place where we presently are. When we are looking for our purpose, we must realize that it might be right in front of us.

What is that thing you do naturally that you don't think of as a gift? My mother tells me I have been counseling people since I was a young girl. It has always come naturally for me. Although I went to college and graduate school to gain a doctorate and to be formally trained as a therapist, it was in my heart all along. That thing which I loved, I naturally moved toward.

For some reason, we tend to believe our natural gift must be hard. That's a cognitive distortion. If it is natural for us, it should not be difficult. What is easy for me, however, would not be easy for a person who did not have the natural gift of therapy.

Look at your gift and dare to wonder, *How can I use this in the service of others?* Be you, but don't compare yourself to others. You have different gifts; therefore, you should have different visions for your life. If you compare, then you lose your joy. Benjamin Franklin once said that comparison is the thief of all joy. Each of our gifts is different, so comparing them is counterproductive.

Sometimes we discount our purpose because others tell us it is not our gift. I once knew a fellow who was told by his high school and then college football coach that he was not gifted as a football player. They advised him to get a desk job. But he continued to play and build his skills for one reason—he loved the game.

After graduating from college and seeing many of his teammates get drafted to play professional football, he had to try out as a free agent. He ended up walking on with a professional football team and played for many years at this level. He was always grateful that he did not listen to those who told him to get a desk job.

No one knows you better than you. Listen to the truth your gut is telling you. Sometimes we don't get paid for our gifts at first. We have to be willing to give our gift for free. We have to be willing to serve others before we can monetize our gift sometimes. As we serve, our platform grows.

When we do the right thing for the right reasons, success follows. We grow. Money is not everything. Doing what you love and following your dreams can lead to amazing things. It can lead to your truth and living your most authentic life.

CONNECT & ENGAGE: *Think about a limiting belief that is holding you back. Where do you need to recognize it for the lie that it is and replace it with a new truth?*

CHAPTER SIXTEEN

Confidence

*Be who you are and say what you feel because those who
mind don't matter and those who matter don't mind.*

—BERNARD BARUCH

MY PERSONAL RESEARCH supports that those who
are less active on social media and the internet, in general,
seem to have more confidence than those who are spending
more than three hours a day online. There are many different
possible reasons for this. It could be that too much social
media comparison lowers self-esteem. Or it could be the
chemical component of too much internet surfing. Maybe
it is the sedentary lifestyle and not enough exercise?

We know that a lot of how we feel about ourselves when
we are by ourselves is a determining factor for happiness.
Happy people are more confident. So it is significant to
pay attention to how we feel when we are alone, when no

one is observing us, and no one is present to be critical or encouraging. The person that we are when we are alone, and how we nurture and grow that person, determines our confidence level.

This is especially true of younger people. In teens, confidence levels plummet when they compare themselves to others on social media. They don't yet have the discernment skills to logically understand that everything they see on the internet is not real.

Humans are wired to compare socially, and now more than ever, the opportunities to do so abound. The judgements we make when viewing others' social media can send our moods plummeting. Research is teaching us that social media in general is making us feel bad. An increased use of social media is causing more depression, anxiety, and envy. Using social media as a habit to avoid bad feelings often just causes us to feel worse. It can be a vicious cycle.

We have so many opportunities for online distraction that it takes us away from experiencing real life and building self-worth. Gaming online and winning is not the same as going to a soccer match and playing in person. You miss out on the subtleties of interacting with real people and working together as a team. Imagine the self-growth when we join friends for a game of doubles and we interact. Think about how it feels to enjoy the eye contact and the congratulatory embraces afterwards. This is what we are wired for as human: in-person connections. Building in-person relationships builds the kind of security that reduces anxiety and depression.

Every day we have a choice to get online or go deeper into ourselves and live with true meaning. This means finding more fulfilling tasks and engaging life with more purpose. This builds our strength emotionally and increases our confidence. Avoiding our personal growth weakens us. We can't just stay the same; we must always be moving forward and towards our purpose and life goals. This is a sure way to build confidence. The opposite is stagnancy and atrophy.

Life is meant to be lived. As we live and experience life, our confidence grows. On the other hand, we don't build much self-esteem sitting in front of our computers or on our smartphones. Personally, after a full day of writing on my laptop, I might go out for dinner with friends and find it really hard to connect with them at first. I feel very disconnected and distracted until I settle in and make a genuine effort to have in-person conversations. There is nothing so real in life as the things you do in person and the experiences you have in the world, living your life in the community of others.

Achievement is the surest way to confidence. Achievement is knowing that you have worked hard and done the very best that is within you to do. Success when others praise you is a great feeling, but it is nothing like the self-validation of hard-earned achievement. Always make your goal *achievement* rather than *success*. Achievement removes the insecurities and lets you enjoy being you.

To be confident, the first thing you must do is fall in love with yourself. There is no other way to feel good about you than to love yourself. When you love yourself first, all of the other things work out and begin to fall into place. Make peace with yourself, and your confidence soars. The most profound relationship you will ever have is the one with yourself.

When we love ourselves unconditionally, we are free to live our lives in person. It frees us to be connected and engaged in every aspect of our lives. Listen to yourself and not the others. Decide what you will need to build your self-esteem and then do it. It is hard to fight the enemy that camps in your head and says, *You are not enough*. Be sure to create life experiences every day that build your self-confidence rather than tear it down.

There is much that goes into building a child's confidence. Childhood is when confidence building begins. Confidence doesn't just happen by accident; it has to be built. And like all things, this is not always an equal building process. Not all kids are provided the same good foundations built by healthy parents. Parenting today requires much greater involvement than it used to. We not only have to worry about the physical world, but we also have the virtual world, which influences our kids just as greatly, if not more so.

Just as you would not allow a sexual predator to babysit your kids or come into your home with them, you should be mindful of what they are subjected to on their digital devices. The virtual world can be a very scary world. Our kids need our guidance to understand all that is out there and learn how to properly manage and process it.

Social media alone can be full of negativity and bullying, and this can take a toll on a kid's confidence. Confidence is critical to set our kids up for a healthy future. I know parents have too many things to do and too little time. I get it, because I've been there. But if that's the case, and you can't make monitoring your kids' digital device usage a priority, then simply don't allow the devices to be a part of their life. We have to make choices sometimes, and that may be the wisest one you can make.

Unmonitored social media usage can be incredibly destructive for kids. Heck, it can be destructive for adults. You wouldn't give your kids the keys to the liquor cabinet, so don't hand them a digital device without monitoring their use of it. Be sure to have boundaries and consequences in place. It's the responsible thing to do.

I see so many parents begin using digital devices to babysit their kids. They are at dinner, and they pull out the digital babysitter in the form of an iPad for their kids to be amused so they can visit with friends. Wouldn't it be better to instead use this time as a learning opportunity to build your kids' confidence and teach them to interact properly with real people in a social setting?

The other day I was having a mani/pedi, and a young mom came in with her five-year-old daughter. She put the girl in the spa chair next to her while she was getting her pedi. She reached into her bag and handed her a digital device with a game she could play.

It was noisy and sounded pretty violent, so the mom started getting some stares. By way of explanation and to no one in particular, she finally said, "We don't allow her to watch TV because there is so much violence. But this is a special treat for her while I get my pedicure."

Clearly, this mom's priorities were a little confused. She was willing to make an exception for the violence if it benefited herself. I get that young moms need some treats and time for themselves. But perhaps you should leave the child at home with a sitter who will run and play outside with her. Then it is a win/win.

If money for a sitter and a pedicure is the issue, I get that too. Then wait until Daddy comes home at night from work to watch her, and then go for the pedi alone. That way, no extra money is needed for a sitter. No dad to come home at night? Then find a friend you trust that you can swap sitter time with and treat each other every now and then to watching each other's kids while you do something for yourself.

There is always a way around things that are not good choices.

A real confidence builder for our kids is understanding what they are capable of and then setting an expectation for them at that level. As a society, we have moved away from chores and setting responsibilities for our kids. We have a generation with no idea of how to make a bed or wash clothes.

Did you know that just having your child make their bed at the beginning of the day can lift their confidence level *and* improve their day? If they begin their day with this goal and achieve it before ever leaving the house, they've started their day with a win. Kids feel really good when they achieve things. That's why teachers used to give out gold stars. Productivity is one of the most rewarding human experiences. Give this gift to your kids. Give your kids chores—expect them to do them—then watch their confidence level soar. Help your kids become aware of the individual resources they possess, and teach them how to use them.

My daughter's job during high school was to mow the lawn. Yes, I know, today that sounds like child abuse. But she was saving for a car, and this was a chore she did to make money to buy her first vehicle.

One day when I returned home from work, she was sitting on the front porch eating ice cream. She had a male friend from school mowing the front yard. When I walked up and asked her what was going on, she let me know that I did not say she had to do it personally, just that she was responsible for seeing that it was done. Well, I couldn't argue with that. She was getting the job done, and I was pretty proud of her entrepreneurial spirit at such an early age. She got the job done in her own way. In this case by using her brain, not her brawn. There are many healthy methods to get the job done. She figured that out at an early age.

Being a good parent and building your kids' confidence often requires balance between freedom and direction. Sometimes parents can allow their kids to make choices and things work out. Other times they will simply have to tell the kids the way it will be. Deciding when to use each method requires judgement. Knowing when to do what is tricky sometimes, and certainly not easy.

Varying chores for kids helps them learn different skills and build competence across many areas. Check your rules from time to time to see if they need to be adjusted as your kids age. For example, a seven o'clock bedtime is great for a toddler, but when the child is in middle school, a later time will be more appropriate. Wise parents notice as the kids age that they have new capabilities. For example, when the child begins to toddle, let them walk. Carrying them everywhere doesn't build their walking skills. The creative parent utilizes their kids' skills to contribute to the family. This builds confidence in the child and benefits the family system. You can't expect your kids to be responsible if you don't give them responsibilities.

All kids need to feel they are contributing and matter to the family system. A kid that isolates in their room scrolling, liking, and gaming is not contributing and not happy. We know from recent research that these kids are depressed. It's up to us to shift parenting expectations for our kids. Alone time is good, but too much of anything is a bad thing.

Balance is key here as in all things. A balance of socializing, working, and playing builds confidence. Family meals together strengthen each member individually. Kids who eat one meal a day with their family and no digital devices are happier and emotionally healthier.

Working with many families that do not come together for a meal each day, it is always my first priority to make this happen. The benefits show up soon after this becomes a habit. Make mealtime at your home a safe place and a place for each person to show up and be valued by the other.

Home should be a safe place to land at the end of the day. With childless couples, reconcile your day together around a meal each day with no digital interruptions. Be present and watch the relationship grow. It sounds simple, but it pays dividends.

CONNECT & ENGAGE: *Think about your confidence level and, if you are a parent, the confidence level of your kids. Where do you need to begin building confidence? What are some small steps to build quick wins?*

CHAPTER SEVENTEEN

Vulnerability

Vulnerability is about showing up and being seen.

—**BRENÉ BROWN**

LEAVING DISTRACTIONS BEHIND and tuning into what you know to be true requires courage. It can be confrontational to listen to what you know, step into your own reality, listen to your intuition, and act from the truths you find there. But if you are unwilling to do so, you'll never grow.

We spend our days searching for answers to the questions we have. But all too often, we don't look inside for our answers. Instead, we try to find them in the external world. We make excuses for what we don't really want to know, and we pass up opportunities for learning. We allow distractions to take us away from our internal compass because, let's face it, the truth can hurt. When we do this, we lose the ability to trust ourselves.

We can learn to trust ourselves again by intentionally listening and being in constant touch with ourselves. It requires vulnerability. It is difficult at first because we have become so good at self-betrayal and abandonment of self. Every day we experience the external programming that helps us along with self-abandonment. So we must begin this journey into finding us again.

The greatest gift to ourselves is the gift of self. Yes, others are significant in our lives, but we must be at home with us and in touch with our true selves. You must be the expert on you.

Part of knowing ourselves includes noticing the parts of us we reject or dislike. We can also take notice of unhealthy patterns in our lives that continue to show up. What part of your life do you need to change so this pattern changes? What story are you telling yourself that is keeping this change from occurring?

For me, an unhealthy pattern I have worked hard to change is sitting down to write this book and, ironically, not getting distracted by social media. I love to see what my friends are doing and stay up to date on what I am missing out on while I am home writing my book. The struggle is when the snowball starts down the hill. It builds speed until my allotted writing time has been spent scrolling and liking.

If I'm not careful, I can be hard on myself because I know I've let myself down.

I am really working hard on my boundaries because I know myself well. I know what I have to do to keep up my work ethic to finish this book. I have chosen to consciously structure my life to suit my goals and focus on the important things. I've chosen to eliminate the things I don't wish to be distracted from.

I don't want to walk through my life asleep. So to avoid this, I daily question my thoughts and feelings. This practice helps me understand my choices and reactions.

It helps me be present in my own life. It helps me understand what is mine and what is someone else's. So much of what we do and think has been handed down to us and imprinted on us by others. We must be intentional and vulnerable to understand what part is ours.

The opposite of being vulnerable is being controlling. It is not possible to love a person and seek to control them. When we are trying to control another, we are simply in denial about our own wounds. Quite often, control is a cover from fear of abandonment. Most often this is part of a story from our childhood, maybe a story that we have not processed and healed.

Until we heal, we will control others rather than being vulnerable ourselves. The caveat here is that when we control, we risk inciting the very behavior we most fear—abandonment. Control becomes an obstacle for things like intimacy and connection. It destroys relationships. When we heal our wounds, we no longer need control as a defense mechanism. Being vulnerable means we don't feel the need to rationalize our behavior and blame others for it. When we move towards the pain and heal it, we are connecting with our wounded parts and investing in ourselves.

Push back against the obstacles and old hurts that keep you from receiving love and letting go of control. Be vulnerable. Do the work on self, and your connections will improve and life in general will be much richer.

CONNECT & ENGAGE: *Think about your response to vulnerability. How easy is it for you to let your guard down and authentically be yourself? How might you begin to be more vulnerable with yourself and with others?*

CHAPTER EIGHTEEN

Connecting with Your Inner Child

It sounds corny, but I have promised my inner child that never again will I ever abandon myself for anyone or anything else again.

—WYNONNA JUDD

WE ALL HAVE an inner child in us, but do we recognize this child and pay attention to it? Do we check in with her to see how she is feeling? That little girl or boy has been with you since the beginning. They've been through it all. They have seen those things that don't go away.

Are you connected with her or him?

When we become adults, we can't become too distracted to check in with our inner child. At first it might be painful, but the better acquainted we become, the easier and more natural the check-ins are. We can learn to reparent this hurt little one inside of us.

If you grew up with little attention, you can love this little one the way she deserved to be loved and make her whole. This can be so healing for the adult version of you. We all had parents that were imperfect, just as we are imperfect. There were some things they were not able to give us because they didn't know how. But now we can learn to give those things to ourselves. This provides powerful healing.

Ask your inner child if she feels heard. How can you be a better listener if she does not? Ask her what she wants you to know. You can reassure her and let her know you want to guide her, and you want her to trust you. All those painful things that happened to her, it wasn't her fault. We will not find peace in the exterior until we have it on the interior. Connecting with our inner child brings great peace.

Think about your reaction to your inner child and how it influences your life. Does it cause you to check out sometimes when things feel too heavy? Do you reach for the remote or the mouse and get lost in your electronics? Would it be different for you if you leaned in to this inner child's nudge to hear what she has to say and be present with her for a bit?

How can you connect with your inner child and celebrate her? The healing that comes with this connection is amazing. Your life will be richer. Take the challenge to nurture this relationship. You can learn to take care of each other and create a life of peace together.

Take the opportunity to reflect, and challenge yourself to gain depth around your inner child. Honor this part of you that you may have ignored or closed off. Transitions are smoother and obstacles are more approachable when you are hand-in-hand with your inner child.

Be willing to put in the serious work now so you lay a foundation forever for this important relationship. Stand shoulder to shoulder with your inner child and do life together.

Be gentle with this little one and grow the compassion she wishes for. Protect her and love her for all time. It's important to remember that the things that challenge you challenge her as well. There is security and safety in doing the hard things together. Bring balance into your life honoring your little child within. Merge the two without either feeling threatened. There is room in your heart for you both.

We must acknowledge our inner child's pain and not shrink it, because when we do, we can't love others well. It stops us from experiencing compassion for others. We must have empathy for self in order to have empathy for others. First, we have to stare down our own hurt in order to hear it when others express theirs. Next, we must be able to change our negative narrative into a positive one.

There is much research around neuroscience and the effects of positivity on our brains. Positive energy is everything. Our childhood may not have been a positive one, but as adults we get to *choose* how we think. We can grow up the negative child in us through our positive messages to self. When our self-talk is negative, we create a negative future.

Those who live the happiest lives act as if they are living in the future and their future is perfect. We can create a perfect future in our lives with our thoughts. When we say *I am*, and follow it with a negative, I attract negative. When I say *I am*, and follow it with a positive, I attract positive.

There is research supporting the fact that happy people live longer, healthier lives, and their relationships are better when they are positive. When we have had a bad experience and we are down and out, we can choose what we believe. We can choose to believe positive things for a better outcome. There is a well-known experiment where people have been given a maple leaf and told to rub it on their arm. Then they are told that it was poison ivy. Not long after, they broke out in a rash. This is how strongly our internal thoughts create our existence. We must believe positive things in order to move into a positive direction.

I have known two men who were told they had terminal cancer and given only a short while to live. The first was my stepfather. He lived for six weeks after his diagnosis. The other was my father. He lived for six days. They each believed the doctor's diagnosis and that there was no hope. They both walked into the doctor's office with hope. They certainly were not expecting a negative diagnosis, yet they both left hopeless and gave up.

Our choices are binary, and they will take us where we think we will go or away from where we wish to go. It is all based on our beliefs. Oprah Winfrey was sexually abused as a child and had a very difficult upbringing. Yet she believed in herself and attained her goals because *she believed she could.* Oprah overcame that rocky childhood, changed the narrative, and believed. And that made all the difference.

Associate with positive people who exude positive energy. Change the narrative on your inner child and see where it takes you. It may not be easy, but that little girl or boy within deserves nothing less.

CONNECT & ENGAGE: *Think about your inner narrative. What wounds are you still struggling to overcome? What words does your inner child need to hear from you today?*

CHAPTER NINETEEN

Respect

This world of ours must avoid becoming a community of dreadful fear and hate, and be, instead, a proud confederate of mutual trust and respect.

—**DWIGHT EISENHOWER**

IT SEEMS WE have all become a little disconnected from respect. Have you experienced being on the other side of a keyboard warrior who says whatever pops into his or her mind? Or maybe you have been one yourself?

It's scary out there in internet land. There's a lot of unkindness. It seems like these days many people go online simply to argue. We don't typically walk up to perfect strangers and tell them our political, religious, and personal beliefs. So why do we feel the need to go online and unload our stuff and then criticize those who don't believe like us? What's up with that?

I've been observing this for a while now, and what I have noticed is this (and oh yes, I know I am going to get some stink eye for saying this): mean people are mean anywhere and always.

They are mean on the internet. They are mean at work. They are just mean everywhere, because that is who they are.

If someone is not living courageously, they will resent your being brave. If someone is angry, they will not like your joy. If someone is dishonest, they will dislike your integrity. It is what it is, so we can't allow ourselves to get bogged down in the quicksand of others' hang-ups.

We have to keep on moving and living and sticking up for all that is right and good in our world. The mean ones are people acting out their wounds. As difficult as it may be, we have to send them compassion because they are hurting.

Anger is a tricky emotion, and many are not taught what to do when this emotion takes over. For example, some people grew up with adults who might have become angry and violent. They may equate anger with danger and therefore they don't allow themselves to go there. We never learn how to manage our anger, because it's not taught in school. Consequently, we have a bunch of grown-ups out there bumping into one another's anger wounds.

When we realize that our anger is a messenger, there is much we can learn from it. What does it feel like, and what is it telling you? Once we get clear on this, then we get to decide what we are going to do with our anger and how we will process it. We become truly brave when we don't use our anger to tear another down.

We can't make others' unkindness about us. We can't dim down because our light hurts someone else's eyes. Give them sunglasses, and be respectful. We can't shrink and become small for others' comfort. The truth is you can never shrink enough for a miserable person.

We all sometimes feel the freedom to say things via the internet that we would never say in an "in person" conversation. We seem to feel safe hiding behind the internet, and that means we have no boundaries.

Good manners are always appropriate—the cyber world is no exception. Being respectful of one another online is more important than ever—especially considering the vast number of people who witness our online behaviors. Typically, we have an audience viewing most of our internet exchanges.

Have you ever witnessed an exchange online where people held different beliefs or opinions, yet were discussing them with mutual respect? It doesn't happen often, but when it does, it's incredibly refreshing.

Being respectful is key to building a positive culture. It shows others that we believe they are people of value. It also shows we value ourselves. It is easy to get so caught up in the culture that we don't notice our own behavior. We simply follow the herd mentality, and it is not until we arrive at an unsavory place that we realize we have made a wrong turn.

Our world has become so focused on *me, me, me* that perhaps there needs to be more thought of others. We need to be a part of one another's lives and create connectedness on a daily basis. We must all get comfortable reaching out to others and lending a hand. Be curious with them about their needs rather than judgmental about their beliefs.

One way to accomplish this is to honor time together with others with no tech connections or distractions. Simply be together and engage and listen. Spend time plugging into one another rather than digital devices.

The big things we do once in a while are not as important as the small things we do daily. We should never stop taking time to learn about ourselves and those we love. We must always be curious and make time to open ourselves up to our loved ones. We should hope that they will also open up with us as we show them how. We must commit to this sort of growth every day as we become a student of our own growth.

We can be intentional about using our energy to understand our life habits and what we are pulled toward. We sometimes get drawn into what is familiar to us. This is due to our core wounds from early life that become our comfort zone. We often find ourselves in dynamics that give us similar pains as we had early in life. If we were disrespected, we move toward this and continue to live this pattern. This is true even if it is not good for us and brings us more pain. Until we become intentional and aware of the need to break the cycle, we will remain there.

We can learn how to be attracted to that which is healthy for us and to respect ourselves by following through and making this happen. When we have clarity around what is not working for us, we can make different choices that end with us feeling honored. As we move away from the distractions that keep us from knowing self, we must move toward internal realities.

Nurture connections that bring real beauty into your life, and appreciate those relationships. You know which ones they are. These soul bonds are not easily found. Cherish them.

When we respect ourselves and each other, it builds community. Respecting ourselves can look many different ways. We might decide that the journey we are on is not the best journey for us, that maybe it is someone else's journey. Respecting yourself means course-correcting. We might need to give up becoming something and instead let go of all of the things that are not really us in our lives. This can lead us to becoming who we are really meant to be, and this is what self-respect looks like. When we become happy and comfortable in our own skin, we are able to respect others and give them compassion more easily.

Sometimes we have to let go of all of the obstacles that block us to the respect we want and deserve. Once we understand what blocks us, we can cut it out and add back what's important. To do this we have to identify our limiting narratives and fears. This helps us know what is keeping us from respecting ourselves and living the life we wish for and dream of. Get curious, and discover what is blocking you from engaging and connecting with yourself and others in a respectful way.

The best way to get respect from others is to respect yourself. You don't have to accept every battle that comes your way. You can simply give the peace sign, smile, and move away. When you trust you with you, you won't feel the need to allow disrespect from others. You will realize that is about them and not you. How are you are abandoning self and allowing disrespect to infect your life and relationships?

When you make a promise to you, follow through with it. Keep the boundaries you set, and don't allow others to disrespect them. Your ability to have good boundaries with other people is directly related to the way you handle the boundaries you keep yourself.

Sometimes the disrespect we project onto others, or they project onto us, is connected to our unmet emotional needs. In order to get the respect you want, you must be able to identify what this looks like for you. If you are stuck in a disrespect cycle with self, then you will find it hard to get the respect you need. Healing is about you owning what you need from self.

We show respect to others when we really hear them and pay attention. To really see another person, we must focus on listening to them. This requires undivided attention. When we make disrespectful assumptions and hold on to them as facts, we can't hear the real message being communicated.

When we are deliberately respectful of others and engaged and connected with them, we build trust and find joy. We show respect when we are connected enough to ask questions. When we seek to understand to be understood, we develop mutual respect. And that's all anyone really wants.

CONNECT & ENGAGE: *Perform a respect evaluation. How respectful are you of others, both in person and online? Where do you need to talk less, listen more, and show more respect with the people in your life?*

CHAPTER TWENTY

Attached

It's not much of a tail, but I'm sort of attached to it.

—WINNIE THE POOH

MY ADORABLE 81-YEAR-OLD mother tumbled down the escalator, backwards and head over heels. I was standing below her in terror witnessing this horrible accident unfold. An angel of a gentleman reached over me to help get her up. He held her up until we could be sure there were no injuries.

That's when I saw it. She had her cell phone in her left hand. She had never let go of it during the entire fall. What could possibly be so important about that digital device that she risk her life to remain connected to her cellphone while all the while, *not connected* to the handrail on the escalator?

Our phones have become our appendages. They are now extensions of us, and most of us are attached to them 24/7.

We truly feel lost without them. If you don't think so, try leaving yours at home and spend just one hour without it. It's like a phantom appendage, something you're constantly reaching for or waiting to vibrate or ring. You feel cut off and alone. But if you can get through that initial uncomfortable period you will feel unbelievably liberated. When you spend time away from your digital device, you will find yourself feeling more connected and engaged. It's a lovely feeling.

It dawned on me one day that we have all become controlled, in a sense, by computers and algorithms. They have become very much like a god to us. For example, have you ever noticed how we react when we drop our phone in water or when our computer suddenly shuts down? We get a little crazy. In many ways, it's similar to a codependent relationship with an abusive partner. We get rewarded… right until the algorithm shifts and then we feel lonely and our mental health suffers. Is that not unbelievable? That multitudes of social media users are having mental health downturns, simply because of the whims of an algorithm? As a practicing therapist, I can honestly say I never would have thought I'd be having these conversations with my clients— conversations about downturns of likes and the resulting depression it caused. These platforms are not designed to promote or foster healthy relationships. Once we come to the realization that social media is rigged against us, then we will hopefully be able to make wiser choices around our relationship with it.

And to take this one step further, it makes perfect (if twisted) sense. Look how important the internet is to our lives. It touches every element of it. Our phones have become an extension of ourselves, attached to all aspects of our being. When our digital devices go down, we go down. Many of us have lovely lives, but we don't notice because we are in head-down, fingers scrolling mode.

I became very cognizant of this when I decided to take a 24-hour digital detox on Thanksgiving Day. It felt so good and also a little nerve racking. That told me I should probably do it more often. It was difficult but, yes, we *can* do hard things, especially when we notice how much better we feel after we have completed something difficult. A digital detox to regain clarity and calm—ah, feels so nice.

We are becoming like robots, and our world is becoming more robotic and influenced by artificial intelligence. I like the meme I saw recently of an international store which uses self-check-out. A shopper being asked to use self check-out said, "No, thank you. I don't work here."

It seems like more and more I am noticing robots take jobs of humans. How will that affect us as humans? Is this the beginning of the Digital Revolution? Robots are becoming so advanced that thought leaders such as Elon Musk and Stephen Hawking have openly expressed concern and opposed using robots for things like war. They fear that at some point the bots will become more powerful than humans, so that we humans will be treated as inferior.

It sounds like a science fiction movie, but let's face it, is it really that far-fetched? Stephen Hawking once said that when AI is capable of intuition, feelings, and self-awareness, this will be a big event—possibly the last big event. Imagine a machine with a conscience. There seem to be humans who don't even have this and we know how well *that* works—not well at all.

But it does seem the genie is out of the bottle, and AI is progressing at a rapid pace. There is no putting it back. It is too late to regulate it. We can only wonder how it will treat humans and if they will be respected. Will these machines be controllable and have a set of ethical laws to regulate them? Is anyone really paying attention? Many are warning of the concerns of unregulated AI, but no one seems to care enough to stop what they are doing and get involved in regulation.

Does this make it dangerous for us? It will soon be outside of human control, and then we can only hope that it will not be used as a weapon against us. We are so busy living our lives that we simply can't be bothered to contemplate how we will be affected by AI until it interrupts our lives in a way that we are rattled, like replacing us in the workplace.

Imagine how plugged-in to Google, Alexa, and Siri we are. Now imagine the possibility of collectively programming AI with our questions. We constantly interact all day long with humans and with machines. It seems this fuels technological growth. But what about personal growth?

It doesn't seem at this point that this move to AI is going to cease. By our constant upgrade to the latest and newest gadget, we are all pushing the progression of faster, better technology. We demand it. As a result, the percentage of intelligence that is not human is growing daily. We constantly project our likes, fears, and hates onto the internet to fuel the online system. We drive it through input from our own limbic systems. I predict that eventually our thumbs won't be able to communicate fast enough, and our computers will be connected to our brains directly.

We can now see our physician online. In fact, in some cases computers have become more efficient at diagnosing disease than humans. The medical community is built around artificial intelligence. I see many clients every day via online communication. Remotely, we make important decisions together that are life changing. It is convenient, I get that, but where is it taking us ultimately? Is it adding to an overall sense of global disconnect?

As I write this, I know that self-driving cars are being perfected, and soon driving will be passé. Honestly, I do look forward to this. But it seems there is a human element that is affected by all that AI is replacing. We will all be affected by AI when it is able to replace almost every element of our lives. It will write books and computer programs. Will AI do construction and art? Already, I have a Roomba vacuum that cleans my floors. And what will I be doing while my robotic vacuum is hard at work? Will I be finding other ways to move my body and feel productive? Or will I be sitting on the sofa, scrolling through social media and eating cookies? America has become one of the most technologically forward countries and also one of the most obese. Perhaps there is a correlation? Is our technology making us unhealthy, fat, and lazy?

Our world will be run by AI and their thinking brains as opposed to our human, feeling brains. There is a toilet that can now give important health information based on urine analysis. Imagine Amazon delivering meds to your door unexpectedly in the afternoon after infection was detected in your morning urine.

We can already control our household appliances from another state. We can order groceries online and have them delivered to our fridge. We have many powers due to AI, but that which gives us power also becomes our god. Think about how the algorithms already run our lives. And oh, the power they have over us! The path we take to our jobs each day is designed by algorithms. The opportunity to biggie-size your meal or get ten cents off at the register—all managed by algorithms.

We like these because they make our lives easier. I don't have to battle traffic to go to the doctor, because I can now see my tele-doc online. I can purchase my seats for the movie online and order snacks to be delivered right to my seats. These are conveniences that most people are not willing to give up. We have made the shift—and we like it. Our lives are easier and more efficient. But is this really better? Don't we need a little effort and friction in our lives to keep us from becoming lazy?

I heard recently on the radio that by 2027, 50% of jobs in the world will be replaced by AI. That is an astonishing number of jobs lost in a very short time. I can't imagine how this might affect our world. Are we prepared for a world where 50% of people are unemployed?

What will those people do all day, considering how we know that our purpose is tied to our emotional well-being?

Supposedly, AI will one day surpass human intelligence. What will that mean for our world? We don't like to think a machine can be smarter than us, so we simply don't acknowledge it. Self-driving cars will soon rule the roads and be 200% safer than people-driven cars, offering a 45% reduction in highway accidents. That seems like a good thing, but I can't help but be curious about what happens to the truck drivers, bus drivers, Uber, Lyft and cab drivers that will be out of work?

AI experts feel the biggest crisis we face is that AI will not be regulated properly, and it will be a serious danger to the public. Some feel AI is to be more dangerous than nuclear weapons. Why then do we regulate nuclear weapons but turn a blind eye to AI? Is someone addressing this behind the curtain like the Wizard of Oz? I wonder if we should all become concerned and interested?

This seems like sci-fi craziness to some, but others believe it is really possible and maybe in our lifetime. There is now a computer that can beat a human at chess. AI can come up with cures for our worst illnesses and make surgery better. AI is amazing and can do fabulous things, and we want more of it. But we must also figure out how to manage it responsibly so we are not managing ourselves right out of a job. What will our lives be like when our purpose is filled by a robot? We know that to be well-adjusted we all need a purpose and a balance of love, work, and play. What will a society with no need for work and no available jobs be like for our psyches?

Some speculate that AI would eliminate the need for humans to work. That sounds boring to me. What sort of things would all of the idle hands get into? Humans need to work. How will they feel about themselves if they are not working? The happiest people I know have a purpose, and many of these people have their purpose linked to their work. Supposedly algorithms are all that is missing and standing in between this bot and reality. AI will be able to do human-level intellectual work millions of times faster than humans. This would obviously bring much change to our society. The best-case scenario will be if this happens and it is regulated. Unregulated it could be a nightmare.

Currently communication between ourselves and our digital devices is slow. This will soon advance, and down the rabbit hole we go. Quite possibly Neurolink by Elon Musk will be the advanced brain-related solving of disorders through implantable brain chips. His goal is to achieve symbiosis with artificial intelligence. His feeling is that in a benign scenario, humans will be left behind, but he is working towards AI that will take us along with it. He considers interfacing the phone with our brains and increasing brain width to create a well-aligned future. As of this writing, he has successfully implanted one human brain with a brain implant from his Neurolink company. At the moment, the person is doing well. Musk hopes to be able to address brain disorders such as Alzheimer's and other dementias (as well as spinal injuries) with this technology. His goal is to redefine the boundaries of human capability. And while many are excited about these new possibilities, many more are understandably concerned. Brain implants have raised a multitude of issues, and offers an opportunity for humanity to draw a line in our physical integration with technology.

Wow! That brings a whole new meaning to being "attached "to our devices. As I sat down to complete this chapter on AI, my daughter texted me pictures of her home she is renovating. She informed me that she had used AI to design the renovations. It really looked great, but I must admit I was a little sad. She and I have always enjoyed breathing new life into old homes and renovating them together. We'd chat over matcha lattes and envision all the changes we would make to rejuvenate those older homes. These projects were about more than just being efficient. They were also about bonding around a common endeavor and enjoying the satisfaction that came when we'd completed a project from start to finish using our brains, hearts and muscles. What amazing timing that she should send this as I was reviewing the ways AI steals our connection and our community. It has the power to take the humanity right out of our lives. We have to be vigilant and mindful to preserve the feeling of taking the long road together, engaging with one another through fun projects and the sweat and tears that come with them. Easier is not always better.

It seems the pace of AI and technology is taking us down a path of silent despair. We are all aware of the problems it brings and yet can't seem to stop the train that is careening down the track, leaving us more disconnected with our emotional lives every day.

We must realize we do still have a choice. We can introduce balance into our digital world. If we take steps to balance digital device usage, then we dictate how we spend our time. We can choose to take the positive and leave the negative aspects behind. We can be in control of how technology impacts our lives no matter what comes our way in the not-too-distant future and what is already here.

CONNECT & ENGAGE: *Spend some time thinking about how AI is present in your life. Is it a good or bad thing? Does it make you lazy and disconnected? Do you sometimes feel that introducing friction into your life through hard work might enrich it and leave you feeling more in charge of your journey?*

CHAPTER TWENTY ONE

Meditation

*The practice of mindfulness is simply being
aware of what is happening right now.*

—UNKNOWN

WHEN YOU GO out in public, almost every person you
see has their head down, phone in hand, headphones on,
talking, texting, or scrolling. I realize progress is coming,
and we have to move with it, but based on the research, I
wonder if this is really the kind of progress we want to make.
We have become so addicted to our digital devices that laws
have been passed in many states about responsible usage.

Our phones have become like extra appendages. Lose it,
and you'll see how lost you feel.

Sometimes, I wonder if the answer we are looking for
might be found if we would just put our devices down, be
still, and listen to what's in our hearts and heads. When we

are not stimulating our brains and instead just sit quietly with things, we might find the answer within ourselves. Eckhart Tolle says, "Wherever you are, be there totally." Are we there totally when we are so distracted by our digital devices that we can't have a brief conversation with a loved one without checking our phones?

Is it any wonder yoga and mindfulness have become so popular in our society in the last decade? We get so overwhelmed with the constant anxiety of our digital devices that it actually becomes a huge relief to simply sit still, listen to ourselves think, and know our hearts. Scrolling and liking all the time sabotages our own inner peace and distracts us from us.

Yesterday, I spent time forest-bathing in a lovely garden in Portland, Oregon. As I sat on a bench with the sounds, smells, and sights of nature vibrantly all around me, I observed people coming and going. Literally every person had their phone in hand. Some sat on nearby benches, necks bowed and fingers scrolling. I wondered, was it intentional, or were they mindlessly scrolling?

Scrolling is the new smoking.

We can't seem to control our phone usage, and it is controlling us. It is escapism in motion. Do we have it in perspective? It seems our quality of life would be much richer if we changed our game and connected in person. We are immersed in so many things online that nothing is getting done well. Rarely do we give things our undivided attention. We confuse productivity with staying busy. We are hardwiring our brains for anxiety.

Here's a list of simple challenges for you:

» Put your phone down and sit with yourself and your thoughts a bit today.

» Listen to the story your child is telling and look into their eyes. Really be with them.

» Stop trying to talk to your spouse while you shop online.

» Don't answer emails while you are spending time with family.

Recondition yourself to give those important people in your life priority over your digital devices. Give yourself priority. Be mindful. How might that change the way you feel about today?

One way for us to have more clarity and focus is to practice meditation. We are constantly distracted by so many things; through meditation, we can become experts at focus. Meditating brings us into a state of now—the present—and it helps us find reality and teaches us to stay in the moment. It helps us embrace and understand what is happening *right now* in our lives. Really, the only thing we for sure have, right now, is the present.

We can't find ourselves on social media. In fact, we lose ourselves there. We compare and judge and move farther away from us, from our true self. Putting your phone aside and focusing on you and your thoughts brings you in closer proximity to you. It brings peace and serenity when you move away from your digital device. The head-down scrolling position, I call it the new global condition, is not as healthy as downward dog.

The good news is that we can train our minds to be connected and engaged through meditation. Can you think of a moment when being connected and engaged is *not* a good thing? To become better connected with others, you will first have to be connected with yourself. You *can* know you. You are worth being known. You are unique, and there is not another like you. It's nice to connect with other people, but we don't have to be them. As you learn to be you, you become more open to connection with others.

When we want to get our bodies in shape, we work out. Our minds can benefit from meditation in exactly the same way. It just takes practice to learn to be connected with ourselves through meditation. Relaxation is also a benefit of meditation and mindfulness. Mindfulness is another word for connection. When we are mindfully meditating, we are reducing the activity in the sympathetic nervous system, and this feels good. It feels peaceful.

There are many benefits of meditation:

» Lower stress levels
» Better blood circulation
» Decreased blood pressure
» Decreased heart rate
» Lower anxiety
» Less cortisol (anxiety chemical)
» Increased feelings of well-being

Through meditation, our immune system benefits, as does our brain health. These are all great byproducts of meditation, but the main objective is simply to *be*. Engage and connect with self. When we meditate, we release our mind from the myriad things it cannot control and our attachment to those things. Internal emotions and external circumstances are two such attachments.

Inner harmony, calm, and peace are the result of connecting with self through meditation.

Learning to focus and manage thoughts can become a beneficial habit of health. Our thoughts are everything. They determine our mood, positive or negative. No one wants to be in a bad mood. Once we learn how to maintain balance in our moods, we connect with peace.

Fatigue and cloudy thinking are part of the stress reaction when we are disconnected. Meditation reduces these feelings and also the inflammation that is caused by stress. Those who suffer with fibromyalgia, PTSD, and other stress-triggered medical conditions have shown positive results with adding meditation to their lives. Stress and anxiety are strongly related, so reduce one, and the other is also lessened.

Many practitioners of mindfulness report an improvement in their feelings about their self-image. In one study, a group of nurses reported experiencing less work stress after implementing meditation into their daily schedules. When the stress at home and work begins to build, it's like a snowball rolling down a hill, picking up speed. Our bodies produce more stress chemicals, and before we know it, we are in full-blown anxiety mode. This can cause depression.

Depression and anxiety are twins and often appear together. Alleviate one source of anxiety, and we can begin to unravel the downhill slide and improve our daily mood and life.

To connect with oneself through meditation is to understand the nature of one's being. If you understand who you are and connect with you, but don't put to use this knowledge of self, then it serves no good purpose in your life. If you can make every effort to practice connecting with self through the daily practice of meditation, then you are serving self in preparation for serving others as well.

The quickest path to well-being is serving others. Serving self is surely a path to self-destruction. When we meditate, we connect with the things we know are not serving us well in our lives. Things like vanity, pride, selfishness and judgement tear us down. Through meditation we can push the delete button to rid ourselves of these elements which are not nourishing us. They become extinct when you do not give them attention in your life. We can enrich our daily lives by thinking *only* on the things that bring us joy and peace. Those things we don't like? Send them good will, and they will cease to infringe upon your thoughts.

When you meditate on the positives and let go of the negatives, you will bless your own life and the lives of those around you. We cannot think of two things at once. If we think on a positive thing, we are not able to entertain the negative or give it space in our mind.

Our actions are an obvious illustration of who we are. Through meditation, we can align our thoughts and actions to be authentic. Consciousness can bring us into true alignment. Meditation is a true positive influence on our lives and comes into our lives to enlarge them rather than restrict. It helps us learn to let go of the things that we will be much happier without. When our consciousness changes, and we are connected and engaged, we are happy to let go of that which does not serve us well. Those things not worth having disappear during the practice of meditation.

All self-understanding that comes through meditation brings about moral and value growth which enriches our lives. Negative thoughts disease our soul, and the behaviors brought forth encroach upon our peace.

Devoting our time to meditation to heal and to overcome unhealthy habits and thoughts is to connect and engage with oneself. We can bring all unconscious thoughts forward through meditation. Your regular actions will demonstrate where you are in your journey into self-discovery through meditation. When the conscious is ready, it will be like a flower opening as we see it unfold into our life habits.

This is a natural growth through meditation. You will be amazed at the pace you grow when you are regularly practicing self-awareness. Trying to make improvements in oneself without meditating and becoming aware will result In repeated failures and frustration. We can move forward into awareness through mediation, and this is an important step in our self-discovery and growth.

We begin to have outer conformity once meditation makes our inner-self conscious. We learn the self-control to master the power over negative behaviors. When we meditate on the positives, we take pause until we rid ourselves of hostile feelings like anger. Practicing anger is to shut ourselves off from positive growth for as long as we remain in this state of mind.

We must constantly watch and meditate to clear our hearts and minds of that which steals our joy and invades our present peace. When we meditate, we learn to pause when a negative thought occurs until we are able to reflexively rid ourselves of such thoughts. Those that have been without for a long time are harder to dislodge, but it is well worth the extra effort.

We must come to terms with all negative thoughts quickly and not allow them room to grow. The cost of carrying the negative is far too great and weighs us down in ways we can't even imagine. Meditation brings to consciousness all that we are not aware of. Sitting quietly with ourselves is to reveal all of self. We are not set free until all negative unconscious is brought forward and processed. By coming to terms with all unconscious negativity, we set our path straight and incur no further resistance.

CONNECT & ENGAGE: *Find a quiet place today in your house or outside. Take a few deep breaths and do your best to process each thought as it comes to you. Take them one by one, and try to determine if it is a thought worth keeping or one that should be replaced. Try to repeat this process at least once a day.*

CHAPTER TWENTY TWO

Generational Connection

*If you are a parent, you have probably already realized
that your children are always watching what you do.
Children watch their parents and emulate their behavior.*

—John Maxwell

WHEN YOU LOOK around, it's easy to see that every generation has been affected by digital devices differently. Yet iGen is the first generation to have grown up constantly with a smartphone. Kids are growing up much differently than generations before. For example, the iGeners are not socializing with others as often and, as a result, they report feeling lonelier. They report not finding life as enjoyable, and therefore they don't feel as beneficial to the world. There is an increase in Major Depressive Disorder, and even more tragically, the suicide rate has doubled in the last decade.

It's not hard to see why. But what to do about it is a more difficult question.

We know that "head down" is the stance for this generation. They spend more time online and on social media than any generation before. They are not sleeping enough and exhibit more risk factors for suicide.

Have you ever noticed that people who are on their phones don't look happy?

We can and must help these kids learn to let their phones be a *tool* to enhance their lives—not a prison that controls their lives. We can teach them to use their phones for a couple of hours every morning and then put it away for the day. We can encourage them to go meet the online friend in person and experience what it feels like to talk face-to-face. They can learn how good it feels to have in-person interaction and to be truly connected and engaged.

Have you ever listened to the monologue that's going on in your head everyday while using digital devices? *Hmm, she's getting married. I wonder if I will be invited. Wow, I am way behind. Everyone in my graduating class already has three kids! She looks amazing in that bikini; I think I'll stay off the beach this year.* When you start to pay attention to your inner dialogue, you begin to understand how powerfully it unconsciously affects your life every day.

The parenting rules for this generation have really changed. Now we have to tell our kids not to send nude pictures to other people! Brush your teeth, eat your veggies, and oh yeah, don't send nudes. Yikes. Digital devices have changed the landscape of parenting. Now we have to help our kids understand how serious things can get if they send nude pics and how the ramifications can follow them for life. Since it's been around all their lives, they have a false feeling of safety and security. Behind a screen, we often forget our behaviors online can be far-reaching.

I challenge you to scroll through Instagram and not see a scantily clad body part. Why, oh why, do we think we have to be naked to get attention? And are we able to discern what sort of attention that will attract and how short-lived it will be? In most cases, this sort of attention leaves us feeling *worse* rather than comforted.

There are no healthy quick fixes online. Helping our kids learn to self-soothe is a gift that they will carry through life. The answers to their biggest questions are not in their phones. When they are down-and-out and in survival mode, they will not feel better after giving too much of themselves away online. Yes, grown-ups are doing this too; that's why we need to model healthy behavior.

Modeling healthy connection for our kids is critical. Helping them view others in a clear way helps them see themselves with clarity. Helping our kids learn a gratitude practice every day is more helpful than teaching them to scroll and like. Teach them ways to be connected to themselves and to connect with others. It is healthy and life-building. Connection is one of our basic needs, so we need to teach our kids to be connected in healthy ways.

Fear is one of our most powerful emotions, and it seems every time I go online, mine is ramped up. There are horror stories everywhere you look online. I am all for being informed, but it seems inciting fear has become a spectator sport. I can't imagine being a child of this generation and every single day being exposed to one frightening thing after another on the internet. I've seen the effect this has on our kids—excessive anxiety. It's no surprise: when you put garbage in, you'll get garbage out.

Think of the effect on your mind when you wake up and scroll through social media, where you see hateful remarks and inappropriate videos of vile crimes or sex acts. What a horrible way to begin your day! The thing we do first in our day sets the tone for the entire day. That's why we must be mindful of how we *start* our day. No wonder there are so many raging drivers and impatient shoppers. Unkindness in general is rampant. It is what kids are exposed to 24/7.

I like this idea: kids check their digital devices in to their parents one hour before bedtime. They get them back two hours after they wake up. This insures good sleep and a peaceful beginning to their day. Most parents have no idea what their kids are doing on their digital devices at night while the parent is sleeping.

The other problem is that anyone can be a "thought leader" now. You simply have to know how to utilize the internet and have a good story to tell. These are the people our kids are getting advice from. These are the people influencing our children's minds. Many have no formal training in psychology, yet they are influencers and offer young people their advice on mental health issues. If they are charismatic and know how to sell their message on video, they are the people influencing what your kids believe. It's scary.

It seems common sense has left the building. We are not thinking logically and asking the right questions. We simply fall under the spell cast by a fast talker and consider their word to be golden. No matter that there is no research to back their statements, and their popularity is driven by likes. With the right amount of persuasiveness, people will buy subscriptions for whatever they are selling. They promise to eliminate your stress or fix your dating life. They are schooled on the hotspots of your pain and know exactly how to push your buttons to get you to push the buy button. Which we do. So do our kids.

We *must* learn to set higher standards for what we put in our brains and what we allow our kids to put in theirs. Once the box is opened, it cannot be closed. We want our kids to hear messages of their self-worth. We want them to develop self-confidence. We want them to live in a world where they feel safe. We want to model safe living for them.

We want them to feel good about who they are and the gifts they bring to the world. Yet if we aren't careful, the device that is in their hands 24/7 can literally wire their brains to make them feel *bad*. What our kids believe about themselves on the inside is what the world will see on the outside. Let's wire them for success and hope. Their minds are so powerful; we can create and destroy them without a thought—simply by controlling *their* thoughts.

The amount of negativity spewed over technology is astounding. Anyone is subject to constant and instant criticism. Few are capable of monitoring their thoughts and filtering their words so they don't come spewing out with venom. We will get what we look for. If you want to see bad in a person, you will see that. What you concentrate on is what you receive from others. We all must learn boundaries and self-control.

There is little of this practiced on social media. It's a free-for-all, and the negativity is killing us. We make assumptions and know so little about others. We say things without regard or factual information to support our comments. We judge others based on what we see in their lives at the moment and not on their entire story. We don't realize that we can only see others from our own level of maturity and perception.

We see so many things on the internet that we want, and it makes us forget what we already have. Instead of being grateful, we are envious. An envious heart takes us to dark places and encourages poor choices. We question the good in others but believe the bad without a thought.

The lesson here is simple: if you want to connect with the next generation, you must learn to see them in a new way. View them through the lens of technology. This new world is affecting them greatly. Be patient, understanding, and present. Be a force for good in their life, and teach them by example how they should live. It will provide a refreshing balance to their digital-heavy lives.

CONNECT & ENGAGE: *Think about the younger generations that you have the capacity to influence. Where might you speak into their lives and show interest in the things they are interested in? How will that be difficult for you, and what can you do to minimize the difficulty?*

CHAPTER TWENTY THREE

Engaged Parenting

It is easier to build strong children than to repair broken men.

—FREDERICK DOUGLASS

IT WOULD NOT be possible to write a book about being connected and engaged and disconnecting from our digital devices without a chapter on parenting the iGeneration. Parenting is a hard enough job; parenting in the digital age takes those challenges to an entirely new level. Make no mistake, we must parent our kids properly and be involved in their digital device usage. Otherwise, their phones become the keepers of their dangerous secrets.

Boundaries are wonderful. We want to teach our kids about them, especially when considering internet safety. I begin my talk to schools and families regarding responsible internet usage by asking them a simple question: *Would you allow a sexual predator into your home for dinner with the family or to babysit the kids while the parents are out for dinner?* Of course, the answer is always no. Then I follow up with this question: Why then do you hand them a digital device where these predators can have unrestricted access to your kids? We don't allow our kids access to the family liquor cabinet, so why do we allow them to have phones 24/7 and not monitor their usage?

Many parents tell me they are concerned with overstepping into their kids' privacy, and that's why they don't look through their phones or monitor them. This is not responsible parenting. A phone is not a journal for your kid's private thoughts. Phones work differently. They give other humans access to your kids online. So until kids are mature and know how to handle these things, we, as parents, must help them with it. We can compassionately help our kids know how to handle digital devices by providing safe boundaries for their usage.

When we don't know what our kids are doing in private on their devices, we can't help them process the things they see there. Some of the things they see they can never be unseen. Kids who I work with often tell me their parents don't know they see nude pictures online. They tell me that they receive them, and they send them. Their parents have zero idea of this because they *don't pay attention* and they don't want to disrupt their kids' privacy. As a therapist, let me be clear: it is necessary. It's responsible. Our kids' brains develop the most between birth and 8th grade; we must do our part to ensure that development is not arrested by digital devices.

In my practice, I see many young men who had access to pornography online as young adolescents between the ages of ten and twelve! Because of this, it is affecting their sexual abilities as grown men in their relationships. They cannot function in a partnership sexually because of their habitual porn viewing and masturbation that started at such a young age.

These graphic pictures children are seeing can never be unseen and will hinder them forever. They can train their brain to pull these images up as needed—even when being intimate with others. Porn is available whenever they want to view it, and now there is even virtual sex available online. It is capturing the attention of our youth very early and affecting them for life. No wonder in-person relationships are failing.

There is an online porn epidemic. Publications such as *TIME Magazine* have recorded the devastation this leaves for our youth. Porn can be a gateway to risky in-person connections. It goes something like this: they start with porn, then move to live-streaming videos and webcams, then finally in-person hookups with prostitutes.

This tragic trail leaves one unable to react physically in a loving relationship. Many men struggle with porn-induced erectile dysfunction. When the brain is saturated with porn, this sabotages the natural sexual responses in monogamous relationships. Porn also objectifies women. In some states, such as Utah, porn is being treated as a public health crisis.

The combination of computers, privacy, sexual pleasure and the brain's behavioral processes lend to porn becoming habitual with all its negative psychological elements. When a twenty-something male expects his new girlfriend to behave in bed like a porn star, it's no surprise that disappointment wreaks havoc on the relationship. Some ten-year old males have told me that when they are left alone, they vacillate between porn and video games all day long.

Many who have successfully quit porn say once it is eliminated, they are able to desire intimacy with a live person and enjoy sex. Often this desire to have sex is the motivator to quit porn. Those who have engaged in excessive online porn say they are no longer stimulated unless sex is like they have viewed in pornography. They describe being desensitized by the 24/7 online porn buffet. Most any fetish wished for can be found online.

Unfortunately, it's young people who are most affected. Forty percent of young men ages 14-17 say they regularly watch porn. Researchers know that porn conditions your brain to need more porn to get aroused. It becomes just as addictive as any drug.

Due to the computer-laden world we live in, it is difficult to avoid. Teenagers taking all of this in while their brains are still developing are quite susceptible—long before they understand the effects porn has on their brains.

If you are a parent and hope to stay connected to your kids, you've got to monitor what they are seeing. It's not an invasion of privacy. It's an inclusion of love and care. It's good parenting. If you want to impact the next generation, it starts with ensuring you are connected to your children and ensuring they are safe and protected.

CONNECT & ENGAGE: *If you are a parent, think about how you monitor what your kids have access to online. How and where do you need to do a better job establishing safeguards around their device usage? Where do you need to set clearer boundaries?*

CHAPTER TWENTY FOUR

The Goal: Well-Being

*There is only time for loving; the good life
is built with good relationships.*

—MARK TWAIN

WELL-BEING IS DEFINED as *the state of being
comfortable, healthy, and happy.* It's a noble goal and one
that every person on earth wants to achieve. Many people
just don't know where to start. I believe being connected
and engaged is the *ultimate* well-being experience. Healthy
connection increases our energy and vitality. This means our
relationships improve and our minds benefit.

Many describe happiness as feeling connected and
engaged. Linking our lives to others creates harmony
and awareness. The openness associated with welcoming
all feelings and integrating them into our lives brings a
sense of well-being and peace. The emotions you focus

on determine the quality of your well-being. Being able to accept all emotions and let them pass through us, not leaving residue, is necessary for well-being. Being connected and part of *we* and not just *me* is a major contributing factor to increasing well-being.

Our thoughts are a large part of our well-being. Our brain is the most important organ in our body, and it controls everything we do. What we tell it is significant and surely affects our well-being. The way we think has a large effect on our lives. Our thoughts are connected to our future and work to create our reality.

We think around 70,000 thoughts a day, but a large percentage of our thoughts are the same as the day before. These same thoughts are what cause the same behaviors and experiences. This is because our personality is created by our personal reality. This is the way we think and feel. If we want a different life for ourselves, then we have to bring our unconscious into our conscious. We have to decide what is working for us and what is not—then we have to let go of that which we know to be problematic for us. Unless we are connected to our true self, we cannot make the needed changes to improve our well-being.

Sometimes we have to completely change how we think.

Our brain is a keeper of our past. All that we have experienced is stored there. If you get up in the morning and follow all of the same habits you always do and go through your day doing you always do, it doesn't challenge your brain. You won't be different the next day simply wishing to be different. You have to think new thoughts and do new things if you want to grow. If we continue to reproduce the same behaviors, we teach our brain to consciously continue the same. By the time we are adults, we live unconsciously with automatic behaviors.

Ninety-five percent of what we do every day is subconscious. Until we learn to make this conscious, there will be no change. We make new connections in our brain as we learn new things. We know that those in the early stages of dementia can offset the disease by learning new tasks. Teaching the brain new ways creates new synaptic connections. Learning to develop a new vision for yourself can actually result in brain health. We can evolve into a healthier brain as we bring ourselves into a new vision. This improves overall well-being.

We often think we are our emotions, and we use our past experiences to define why we are where we are in our lives. Living in the past stops us from creating the kind of new experiences that can change the way we feel. If we continue to live the same life day after day and never change anything, we won't improve our lives and well-being. We must shake things up in our lives to have true change.

If you want to have a different future that is better than your present, you will have to trick your brain into thinking that the good things are going to happen. Then you can move towards them. You can leave poor feelings behind and create new ways to change your mind to visualize positive. We can actually change our bodies by simply changing our thoughts.

If you want to improve your well-being, create a positive vision for yourself and move towards it. Fake it until you make it if you must; truly this will help you gain this wish you have. Use your creative center in your brain, your frontal lobe, to bring your visions into reality. Turn on the part of your brain that is not automatic; wake up the boss of your brain, and tell it what you want and how you want it.

Whenever you change your mind, you make your brain work in a different way. Make it a better way. Your thoughts can inspire and motivate you and improve your sense of well-being. Use your thoughts to leave the past and move into a better future.

What are the things you want to do to gain your vision? Write them down. As you do this, notice what you feel. Use these feelings to move towards even more good feelings.

When you can become conscious of your reality, you can make changes to your life. If old experiences make you feel guilty or sad, leave behind the old experiences. Begin living for the present, and make the present better.

Make different choices.

At first it will not be comfortable because it is all new, and this can be scary. Most people would rather hang onto the old bad rather than feel the new better because the bad is familiar. Don't be most people! When you feel uncomfortable, you know you are making the right moves. Make your body uncomfortable being negative; make it feel comfortable only by being positive. Condition yourself and your body to think differently. It is ok if it doesn't feel right. Right can be wrong, and sameness can mislead you.

Familiar is not always good. Step into the uncertainty and change. The good stuff never occurs in the familiar. Get a clear vision for what you want tomorrow to be for you. Apply this knowledge, and make it work for you. Just thinking about a better tomorrow prepares your mind and body for well-being as your actions align to match your intentions. This is how you get new experiences.

Getting clear on your goal teaches your body a new language. The emotions you have are chemical, so make them positive. Being able to repeat positive experiences can rewire your brain, and it can become the new familiar. This can be a way to become a new you.

Think, then do, then become.

Don't just think about happiness, *become* happy.

Don't just think about joy, *become* joyful.

Use information to transform your life and create a new model for your life. Evolve into this new thing you want to be. Leave the old behind. Yes, it is important to know what you feel, but also to learn to change your feelings to be in line with what you wish for. Your uniqueness depends on how your brain is wired. Since we have all had different experiences, our brains are wired differently. Embrace your differences.

Our environment is what separates us all. People, places, and things are what comprise our environment. So if we have the same brain, but we have different environments, then we are different because of the environment. Our environments connect us and bond us. Our traditions connect us and bring well-being. Our environments are changing quickly as we move toward a global culture. Clarity around our purpose brings vision for our environments.

If we wait to *feel* happy to *be* happy, we are not living in the present. To find our success, we must envision abundance and move toward it. Most people wait for the external before they change their internal. If this never happens (which is often the case), then they never move forward.

We can't wait for our reward to feel motivated. We have to feel motivated first and move toward that vision. Our vision fades otherwise. Think about Nelson Mandela. He had a vision of fairness, and he motivated others based on this vision. He did not *wait* for it to happen; he moved toward it and took others on this same journey with him. You have the same capabilities as Nelson Mandela and can live in the future rather than the past.

Cultivate new internal feelings and emotions rather than allowing past traumas and experiences to hold you hostage in yesterday. Instill change in your world that is bigger than you. Think of some of the most successful world changers; they had a vision and held on to it and moved toward it. People who make a difference in our world have visions bigger than them. They are self-actualizers. These people do it because they *say* they are going to do it. They follow their internal convictions, and this motivates them to action. This helps them build a good life, a life of well-being.

We often operate under the impression that being rich, famous, and hard-working will bring us well-being. It's not true. When speaking with older people about their past and what has brought them well-being, they'll say that connection with others is the most significant indicator of well-being. Most find this more significant than wealth and career.

The lessons we have learned from living our best lives are about the power of relationships and connections. At the end of the day, the people who live the best lives are the people who lean into building relationships and community.

They replace screen time with live people.

They build relationships.

They take ownership of their life.

This can be you. It's time to connect and engage.

Dr. Lori

Helping businesses and families manage technology distractions and establish real connections

---

WWW.DRLORIWHATLEY.COM

ABOUT THE AUTHOR

DR LORI WHATLEY is a Clinical Psychologist who specializes in the effects of excessive digital device usage on individuals , family systems and culture in general. With over 25 years' experience, Dr Lori has worked with thousands of clients both international and domestic. Dr Lori has found success working with CEOs , leaders of multi-billion dollar industries , clergy, professional athletes and, most importantly, moms and dads striving to find balance for their families in an everchanging society geared toward constant digital device usage.

As a thought leader in her field, she has been featured in *People*, *Oprah*, Yahoo News, ESPN, *Psychology Today*, Fox News, the *Wall Street Journal*, IHeart Radio and many more publications and podcasts. In addition to working one-on-one with clients in her private practice, Dr Lori also manages a busy public speaking calendar as well as frequent guest appearances on podcasts.

Through countless hours of research, Dr Lori is committed to helping society overcome the endless problems created by digital device usage. She is convinced that the rise in mental health issues can be lessened through more in-person human connection and less online time. While she believes there is a useful place for technology in our world , there should also be balance in our digital device usage and in-person engagement to enjoy the optimal healthy lifestyle.

ACKNOWLEDGEMENTS

I OWE AN incredible debt of gratitude to many people who have contributed to my life the experiences and knowledge I've shared throughout this book.

To the kind people at Hatherleigh Press who walked me through this process and shared in my wish to educate our world about the pitfalls of being too wrapped up in our digital devices and disconnected from in person human interactions. Thank you Hannah, Ryan, Ryan and Andrew.

To my faithful clients who have taught me more than I could ever learn in a classroom, I am grateful and I thank you for the great honor of being a part of your courageous journeys.

To my dear friends and family, thank you for encouraging me to write. Thank you for the gentle nudges toward my goals and cheering me on . You are truly the wind beneath my wings.

ENDNOTES

Introduction

[1]   Joanna Stern, "Cellphone Users Check Phones 150x/ Day and Other Internet Fun Facts," *ABC News* May 29, 2013, accessed December 5, 2019. https://abcnews.go.com/blogs/ technology/2013/05/cellphone-users-check-phones-150xdayand-other-internet-fun-facts

[2]   Jacqueline Howard, "Americans devote more than 10 hours a day to screen time, and growing, *CNN Health*, July 29, 2016, accessed on December 5, 2019, www.cnn.com/2016/06/30/ health/americans-screentime

[3]   Jean M. Twenge, "More Time on Technology, Less Happiness? Associations Between Digital-Media Use and Psychological Well-Being," *Sage Journals*, May 22, 2019, accessed December 5, 2019. https://journals.sagepub.com/ doi/10.1177/0963721419838244

[4]   Marc Nathanson & Matthew Furhman, "Football coach hailed as hero for disarming gunman outside Oregon high school," *ABC News*, May 18, 2019, accessed December 5, 2019, https:// abcnews.go.com/US/school-staffer-disarms-gunman-oregonhigh-school/story?id=63111074

Sleep Smart

[5]   William Shakespeare, "Macbeth," Act II, sc. 2, 1606.

[6]   Preventive Medicine Reports, Volume 12, Pages 271-283, December 2018, accessed December 5, 2019. https://www. sciencedirect.com/journal/preventive-medicine-reports/vol/12/ suppl/C

[7]   M. Nagai, S. Hoshide, and K. Kario, "Sleep duration as a risk factor for cardiovascular disease- a review of the recent literature," (1):54-61. doi: 10.2174/157340310790231635, *NCBI*, February 6, 2010, accessed December 5, 2019. https://www.ncbi.nlm.nih.gov/pubmed/21286279

[8]   S. Cohen e al., "Sleep habits and susceptibility to the common cold." 62-7. doi:10.1001/archinternmed.2008.505, *NCBI*, January 12, 2009, accessed December 5, 2019. 62-7. https://www.ncbi.nlm.nih.gov/pubmed/19139325.

[9]   A. Kuerbis et al., "Substance abuse among older adults." 30(3):629-654 DOI:10.1016/j.cger.2014.04.008, *Europe PMC*, June 11, 2014, accessed December 5, 2017. http://europepmc.org/article/PMC/4146436

Focus

[10]   Kevin, Mcspadden. "You now have a shorter attention span than a goldfish." *Time*. May 14, 2015, accessed December 5, 2019. https://time.com/3858309/attention-spans-goldfish/

[11]   Roy Greenslade, "Ap Reporters Told to File Shorter Stories." *The Guardian*. May 2014. accessed December 5, 2019. https://www.theguardian.com/media/greenslade/2014/may/13/associated-press-us-press-publishing

Disconnection

[12]   Rachel Ehmke, "How Using Social Media Affects Teenagers," *Child Mind* Institute, accessed December 5, 2019. https://childmind.org/article/how-using-social-media-affectsteenagers/

Empathy

[13]   Terry Christopher & Jeff Cain. "The Emerging Issue of Digital Empathy." *American Journal of Pharmaceutical Education*: Volume 80, Issue 4, Article 58, 2016, accessed December 5, 2019, https://doi.org/10.5688/ajpe80458

Stress

[14]   Rachel Ehmke, "How Using Social Media Affects Teenagers," Child Mind Institute, accessed December 5, 2019. https://childmind.org/article/how-using-social-media-affectsteenagers/

Community

[15]   Catherine Paddock, "Loneliness Tied To Higher Risk Of Dementia," *Medical News Today*, October 31, 2018, accessed December 5, 2019. https://www.medicalnewstoday.com/articles/323535.php#1

Confidence

[16]   Denis Campbell, "Depression in Girls Linked to Higher Use of Social Media," The Guardian, January 3, 2019, accessed December 5, 2019. https://www.theguardian.com/society/2019/jan/04/depression-in-girls-linked to higher use of social media

Attached

[17]   Yuval Harari, *21 Lessons for the 21st Century*, 1st ed., (New York, New York: Spiegal and Grau, 2018.)

Meditation

[18]   H. Williams, L.A. Simmons, P. Tanabe, "Mindfulness-Based Stress Reduction in Advanced Nursing Practice: A Nonpharmacologic Approach to Health Promotion, Chronic Disease Management, and Symptom Control," (3):247-59. doi: 10.1177/0898010115569349, *NCBI*, February 11, 2015, accessed December 5, 2019, https://www.ncbi.nlm.nih.gov/pubmed/25673578

Engaged Parenting

[19]   Belinda Luscombe, "Porn and the Threat to Virility," *TIME*, March 31, 2016, accessed December 5, 2019, https://time.com/magazine/us/4277492/april-11th-2016-vol-187-no-13-u-s/

[20]   Patrick J Carnes Ph.D, *Out of the Shadows*, 3rd ed., (Center City, Minnesota: Hazelden Publishing, 2001).